Goatkeeping 101

Third Edition

Sensible information from people who know and raise goats

THIS BOOK IS DEDICATED TO ALL THOSE WONDERFUL GOATS WHO HAVE PASSED THROUGH OUR LIVES IN THE PAST THIRTY-FIVE YEARS

Caprine Supply
Goatkeeping 101: sensible information from people who know and raise goats/ Caprine Supply

ISBN 978-0-9841656-0-5

Published by:
Caprine Supply
P. O. Box Y, 33001 W. 83rd St.
DeSoto, KS 66018

Acknowledgments

A number of people have contributed to the three editions of *Goatkeeping 101,* and we wish to acknowledge their help and thank them sincerely. These people include:

Dave Thompson, publisher of the *Dairy Goat Journal*
Jeff Klein, publisher of *United Caprine News*
Ellen Herman, formerly of *Goat Tracks: Journal of the Working Goat*
Terry Hankins, publisher of the *Goat Rancher*
Robert Spencer, Alabama Cooperative Extension System
The American Dairy Goat Association
Gail Bowman, author of *Raising Meat Goats for Profit*
David Millison of the Nigerian Dwarf Goat Association
Maggie Leman of the National Pygmy Goat Association
Pat Showalter of the Kinder Goat Breeders Association
Janet Hanus, Pygora Breeders' Association
Roylyn Coufal, former publisher of *The Goat Magazine*
Linda Fox, *CashMirror* Publications
Steve and Sylvia Tomlinson, breeders of Kiko Goats
McCrory Mfg. Co., makers of Parmak fencers

and the following authors of included articles:

Drs. Brian McOnie and Murray Flock, Cathie Keblinger, Jeff Klein, Pat King, J. Marcos Fernandez, Dr. Stuart Southwell, B.V.Sc, M.R.C.V.S., Graham Culliford, Nat Adams, DVM, author of the coccidia and worming charts, Mary Wolf, Rex and Terri Summerfield, Rusty Fleming, and Judy Kapture, who wrote a number of the articles appearing in the early Caprine Supply catalogs.

The illustrations represent the work of Jude Fulkerson, Sara Gatling-Hones, Christina Cassady, and Denice Munchrath, and we thank them for their contributions.

We also cannot forget to thank our great staff at Caprine Supply who were instrumental in getting this project off the ground and, through the year, have contributed suggestions for content and corrections. Also, in addition to their regular daily duties, they bravely suffered through the reading of draft manuscripts.

Table of Contents

Table of Contents

Introduction

We are pleased to tell you that over the years, *Goatkeeping 101,* originally published in 1998, has sold more copies than all of Caprine Supply's books combined. In 1999, we revised it, clearing up small inaccuracies and typos. We thought that was it. However, when we needed to order a new print run in 2009, it quickly became evident that this book was quite out of date and needed a complete revision. This Third Edition is a result of these efforts.

Since, 1998, there have been a number of changes in the goat world. The American Dairy Goat Association now registers two additional breeds: Nigerian Dwarfs and Sables. We have included these breeds and their accompanying registries here. Also we have added web sites for as many associations, registries, clubs, magazines, laboratories, and the like as we could find. You'll find a new list of goat-related websites added to the Resources section. We've added sections on milking machines and embryo transfer. Also included are a current article on meat goats and 2008 DHI data. The Resources section has been updated so it is current for 2009.

We hope you find this Third Edition helpful and fun to read.

In order to understand how *Goatkeeping 101* came to be, please read the original "Introduction" below:

"As the years have passed, the Caprine Supply catalog has gotten larger and larger. Each year we have tried to bring you new items we feel will make a positive addition in your goatkeeping life. We have also tried to bring you articles of interest and hints, especially for those people just beginning with goats. However, as more items were added, space requirements forced us to "retire" certain articles from the catalog. At the beginning of our twentieth year, we wanted to bring back all the information that had been retired from the catalog over those years.

This book, *Goatkeeping 101,* began as a collection of those articles and hints, but it soon became clear that with only those articles the book would be incomplete and, therefore, not as useful as a more comprehensive collection would be, so we solicited materials from our leading goat journals, associations, and goatkeepers themselves.

While we originally intended to publish a small, pocket-sized book, the number of contributions has been overwhelming, and we collected more material than we could fit in one book. We now look at *Goatkeeping 101* as a "work in progress." We hope that as you read it, you will note areas that we can expand on or new topics to include, and will let us know. We plan to revise our book, with your help, on a regular basis, so it can reflect current practices and new areas of interest.

The book is divided into sections such as "Starting Right," 'General Management," "Housing, Feeding, and Fencing," and "Goat Products," to name just a few. Most sections begin with an overview, and then contain specific articles, hints sent in by our customers, and light-hearted general interest articles originally appearing in the *Dairy Goat Journal.* We are especially appreciative to Dave Thompson, its publisher, for allowing us to reprint these.

While much of the information (as the name *Goatkeeping 101* implies) is appropriate for beginning goatkeepers, we think there is enough new information, especially about various breeds of goats and the excellent worming and coccidia charts, to be helpful to the most experienced goatkeeper.

The information in *Goatkeeping 101,* including the articles, hints, charts, and checklists, should be used as advice only. While we have tried to ensure that the information is correct and that the practices are good ones that have worked for goat owners, every herd is individual, and you are most knowledgeable about what will work for your herd. Use your own judgment as you read and make decisions about whether to try out our advice. Remember: there is no one right way to do things.

As always, be sure to check with your veterinarian about any questions you have about your goats' health. The information here is meant to be a guide only. You and your vet should act based on the conditions present in your herd.

Caprine Supply has always been your "Goat supply connection—bringing you sensible, quality equipment from experienced goatkeepers." In the same way, *Goatkeeping 101* brings you sensible information from experienced goatkeepers, people who really know goats."

Starting Right with Goats: An Overview

If you were to take an informal poll asking how goat owners first decided to raise goats, it soon would become obvious that this decision often is not a conscious one. While answers vary, most people do not get up one morning and simply decide that they can not live without owning goats for the sake of the goats themselves.

The more usual routes to goat ownership may be a family member's cow milk intolerance (that was a common reason before the emergence of soy-based milk formulas), an overgrown pasture, lonely horses, a pathetic looking creature in a neighbor's field, or a longing to get back to a more simple way of life.

What is clear is that most of us do not know much about goats when we get them, and more important, do not know in advance the demands that goat owning will make on us. Such knowledge comes incrementally, over time. Unfortunately, goats tend to multiply exponentially over less time!

Not only do we usually not know anything about animal husbandry, but we are often even less aware of the differences between the kinds of goats available for our consideration.

To start out right, you need to learn as much as you can before you get serious about goat owning. And never stop learning, either from others around you with like interests or from experts with knowledge far beyond yours.

You must also make a number of decisions that include (but are not limited to):

Deciding why do you want goats in the first place. The answer to this question can lead you in a number of directions: for milk, for meat, for fiber, or for companionship.

If you decide on milk goats, you must remember that this decision commits you to milking goats, twice a day, pretty much year round.

Non-milking goats may mean less time commitment, but may mean an equal financial outlay for stock.

If you have children who wish to be involved in 4-H or FFA, you can choose between dairy goats and meat goats, but you may need to purchase registered or recorded stock. Before making the choice between dairy and meat goats, be sure your local 4-H or FFA program includes an ongoing project.

Starting Right with Goats: An Overview

If you do not want your goats to be a financial drain, you may need to develop some marketable product(s). This could include milk or other milk products like cheese, yogurt, or soap, meat, hides, or breeding stock.

Once you have made a decision about whether you want milking or non-milking goats, the tendency is to jump right in and buy whatever "looks good," or is priced right.

A little planning can save you lots of headaches and money. Most goat owners will be willing to help you choose animals that most suit your needs. For example, if you are getting goats for a young child's 4-H project, it would be unwise to start with grown Saanen or other large does. They would be too much for a young child to handle. Young animals of a breed that your child finds appealing and that are popular in your area would make a better choice in the beginning. If you need high fat milk for the goat products you plan to develop, then you should consider Nubians, not Toggenburgs.

If you plan to breed your goats (even if you do not plan to milk them, you will probably want to breed them if only for replacement stock), you will need to consider whether there is breeding stock available locally for you to breed your goats to.

If you plan to show your goats, you should consider if there are any other breeders of either the type or breed of your goats within acceptable driving distance. Most shows require specific numbers in order to be official.

If you plan to market goat products, you are taking on an extremely complicated task. Not only are you trying to raise goats, but you are also trying to establish a business for these products. This will necessarily entail market research, knowledge of local, and perhaps national requirements, a processing plant (no matter how small), distribution, and a host of ancillary aids (like lawyers, accountants, health inspectors, computers, etc.). The decision to market goat products should not be made quickly, and it is far more complicated than it appears at first glance.

You will probably go through a few different breeds of goats before you find goats you really want to raise. You also need to consider soundness and correctness of the animals before you make them your own.

If you take the time to gain a little knowledge of what makes a good goat (of any breed) and decide on the type or breed ahead of time, you will probably save later frustration.

Basic Terms and Facts

The word **caprine** is used to describe something as having the characteristics of a goat. Compare it to the words "bovine" and "equine" that are used to describe things as having the characteristics of cows and horses, respectively. **Caprine** is derived from the Latin word "caper" which means goat. The English words "Capricorn" and "capricious" are also derived from the same Latin word.

A female goat is a **doe**. Most people do not use the term "nanny goat."

A male goat is a **buck**. Again, most people do not use the term "billy goat."

Young goats are called **kids**. Growing kids may be called "doelings" or "bucklings."

The **dam** is the mother of a goat.

The **sire** is the father of a goat.

Giving birth is **kidding**.

Giving birth is also called **freshening**. The doe comes into a fresh flow of milk.

To **wean** a kid means to take it off milk and have it eat hay and grain only.

Lactation is the time of milk production. It begins when the doe freshens, and it ends when she dries off.

Milk is measured by weighing. A **pound of milk** is about one pint. A gallon weighs about eight pounds.

The **breeding season** lasts from approximately September until March, depending upon where you live. Does usually come into heat every 21 days.

The **gestation period** of a goat is five months—about 150 days. From one to four kids are born, occasionally quintuplets.

Beards are chin whiskers on goats. Nubians usually do not grow beards.

Wattles are the hair-covered, dangling appendages that some goats have on their necks (or sometimes on their ears!). Usually goats have two wattles, sometimes just one. They appear on most breeds of goat and can be considered decorations since they serve no purpose that we know of.

Becoming an Informed Buyer

When shopping for a car, few of us would ever walk into a showroom and buy a car without asking questions about the standard features, options, test results, etc. However, often when we buy goats, we look them over, pay, and take them home. Knowing what questions to ask and the answers you want to hear will help you become a informed buyer and head off many potential problems.

Before you even ask any questions, take a good look at the goats themselves. Do all of the goats look healthy? Are there any obvious health problems like wheezing, limping, scouring (diarrhea)? Check the overall condition of the goats. Are they thin? fat? just right? Do their coats look healthy? Are their hooves trimmed?

Next, check the environment the goats live in. Is it clean and orderly? Are goats separated by age, grown does from kids? Are bucks separately housed or run with the does? Is the barn clean with not a lot of excess or damp bedding? How does the barn smell? Like ammonia? If you are at all uncomfortable with anything you see there, you should exercise caution when buying stock. The environment goats are kept in will usually be indicative of the care the goats are getting in general.

What breed are they? If you are not shopping for any particular breed, be sure you understand exactly what breed or what crosses the goats you are looking at are or are meant to be. If they are supposed to be purebred or recorded, ask to see the papers. To avoid problems, the registered owner should be the person from whom you are buying the goat (or a member of the family or household). Be sure you understand the difference between an American and a Purebred. This can be important when purchasing dairy goats. Be sure you understand that having papers does not mean that the goat is necessarily a purebred.

Can you see the goat's mother, father, offspring, siblings? There are good reasons to look at relatives. You may be able to see traits, both good and bad, common to a family line. If there are no relatives in the herd, ask why not. Perhaps there were some traits that the owner did not like. Be a bit suspicious if the seller has no relatives, and listen carefully to the reasons why there are not. If you are buying from a prominent breeder, you can ask for the names of others who have purchased related animals. The breeder should be happy to provide you with the names of others who have bought stock both related and unrelated.

For animals without papers, ask **when they were born**?

For does, ask **when did she last kid? How many times has she kidded? Has she ever had any kidding problems? How many kids did she have at each kidding? Were any stillborn, or did any kids die soon after birth?**

For bred does, ask **when was she bred? To whom was she bred?** Be sure you get any necessary breeding certificates if you plan on registering offspring.

For all goats, ask about **any illnesses the goat has had in the past, and any illnesses the herd has experienced.** Be suspicious if the seller says there have never been any health problems. Look carefully for abscesses, swollen knees, scouring, and anything else you do not want to bring back to your goats at home or have to deal with. If this is an expensive purchase, you might ask for a vet check.

For milkers, **ask how much does she milk?** We all tend to exaggerate, so if milk production is of primary importance to you, ask to visit at milking time. Better yet, come twice in a row, so you can get an idea of the total daily production. **Ask to milk the goat yourself.** If you are milking by hand, a difficult milker may not be what you want.

Once you have decided to buy, ask the following:

When was the goat last wormed? what product was used?

What vaccinations has the goat had? when were the vaccinations last given? which products were used?

Is the goat on any medication currently? what and how much?

What is the goat eating? what kind of hay and how much? what kind of grain and how much? any mineral supplements?

Are there any special treats the goat is used to?

Is there a special routine you should know about (for example, the goat always has to be the first goat in to be milked)?

Any bad habits?

What happens if the goat dies within a short time? Do you have any recourse? The seller's response may give you a feel for likely treatment if problems occur with the goat in the future.

Most sellers will be happy to answer all these questions and more. If you encounter evasion or resistance, you might rethink your purchase.

If you cannot make a visit to the seller (having made purchasing arrangements through advertisements, phone, internet. or through a third party), the old saying "buyer beware," should become your mantra. If the seller advertises nationally, you can often contact others who have bought stock previously. If you know someone who lives near the seller, ask him/her to make a visit for you. This person should be willing to make the same kind of evaluation you would if you could be there.

Once the goat comes to its new home, give it some time to get used to its new surroundings. If you can, keep it segregated until you are confident that it will not introduce any disease into your herd and that it is comfortable enough with its new home to take the stress of meeting the herd. Try to give it the same feed it ate at its old home. Make any switch to new feed slowly. Such changes can be stressful and, coupled with the stress of a new environment, may really throw the goat off. If you are buying a milker, be prepared for her not to milk as well as you expect right away. She will need time to get used to everything new around her.

While none of these precautions will guarantee the purchase you are making, becoming an informed buyer will help you avoid many of the potentially disastrous problems that can come from buying the proverbial "pig in a poke."

Breed Standards for Dairy Goats

The American Dairy Goat Association recognizes eight breeds of dairy goats: Alpine, LaMancha, Nigerian Dwarf, Nubian, Oberhasli, Saanen, Sable, and Toggenburg. ADGA registers Purebred and American dairy and records Grade and Experimental goats. They also issue a Certificate of Identification for wethers.

The American Goat Society also recognizes these eight breeds of dairy goats, plus Pygmy goats. AGS registers only Purebred goats of the above breeds.

The following breed standards appear on the American Dairy Goat Association's 2009 website. We thank ADGA for allowing us to reprint them here. The drawings that accompany the breed descriptions are meant to represent a typical goat of the breed. For specific breed requirements, however, refer to the breed standards themselves.

Also, contact national breed clubs for more information and specific characteristics of each breed.

Alpine

http://www.alpinesinternationalclub.com

The Alpine dairy goat is also referred to as the French Alpine, and registration papers for this dairy goat use both designations and they are synonymous.

Alpine Breed Standard

The Alpine dairy goat is a medium to large size animal, alertly graceful, and the only breed with upright ears that offers all colors and combinations of colors giving them distinction and individuality. They are hardy, adaptable animals that thrive in any climate while maintaining good health and excellent production. The hair is medium to short. The face is straight. A Roman nose, Toggenburg color and markings, or all-white is discriminated against.

Alpine colors are described by using the following terms:

Cou blanc (coo blanc)—literally "white neck" white front quarters and black hindquarters with black or gray markings on the head.

Cou clair (coo clair)— literally "clear neck" front quarters are tan, saffron, off-white, or shading to gray with black hindquarters.

Cou noir (coo nwah)—literally "black neck" black front quarters and white hindquarters.

Sundgau (sundgow)—black with white markings such as underbody, facial stripes, etc.

Pied—spotted or mottled.

Chamoisee (shamwahzay)—brown or bay characteristic markings are black face, dorsal stripe, feet and legs, and sometimes a martingale running over the withers and down to the chest. Spelling for male is "chamoise."

Two-tone chamoises—light front quarters with brown or gray hindquarters. This is not a cou blanc or cou clair as these terms are reserved for animals with black hindquarters.

Broken chamoises—a solid chamoisee broken with another color by being banded or splashed, etc.

Any variation in the above patterns broken with white should be described as a broken pattern such as a "broken cou blanc."

LaMancha

http://www.lamanchas.com

The LaMancha goat was developed in the United States. It has excellent dairy temperament and is an all-around sturdy animal that can withstand a great deal of hardship and still produce. Through official testing this breed has established itself in milk production with high butterfat.

The LaMancha face is straight with the ears being the distinctive breed characteristic. There are two types of LaMancha ears. In does one type of ear has no advantage over the other.

1. The "gopher ear" is described as follows: an approximate maximum length of one inch (2.54 cm) but preferably nonexistent and with very little or no cartilage. The end of the ear must be turned up or down. This is the only type of ear which will make bucks eligible for registration.

2. The "elf ear" is described as follows: an approximate maximum length of two inches (5.08 cm) is allowed, the end of the ear must be turned up or turned down and cartilage shaping the small ear is allowed.

Any color or combination of colors is acceptable with no preferences. The hair is short, fine and glossy.

Nigerian Dwarf

http://ndga.org

http://www.andda.org

According to the American Dairy Goat Association's Breed Standards, the Nigerian Dwarf is a miniature breed of dairy goat originating in West Africa and developed in the United States. The balanced proportions of the Nigerian Dwarf give it the appearance of the larger breeds of dairy goats, but does stand no more than 22.5" (57cm) [at the withers] and bucks no more than 23.5" (60cm) [at the wither]. Any color or combination of colors is acceptable.

The medium length ears are erect and alert. The face is either straight or dished, and the hair is short and fine.

Nubian

http://www.i-n-b-a.org

The Nubian is a relatively large, proud, and graceful dairy goat of mixed Asian, African, and European origin, known for high quality, high butterfat milk production.

The head is the distinctive breed characteristic, with the facial profile between the eyes and the muzzle being strongly convex. The ears are long (extending at least one inch [2.54 cm] beyond the muzzle when held flat along the face), wide and pendulous. They lie close to the head at the temple and flare slightly out and well forward at the rounded tip, forming a "bell" shape. The ears are not thick, with the cartilage well defined. The hair is short, fine and glossy.

Any color or colors, solid or patterned, is acceptable.

Oberhasli

http://oberhasli.net

The Oberhasli is a Swiss dairy goat. This breed is a medium size, vigorous and alert in appearance. Its color is chamoisee. Does may be black but chamoisee is preferred. Chamoisee is described as: bay ranging from light to a deep red bay with the latter most desirable. A few white hairs through the coat and about the ears are permitted.

Markings are to be: two black stripes down the face from above each eye to a black muzzle; forehead nearly all black, black stripes from the base of each ear coming to a point just back of the poll and continuing along the neck and back as a dorsal stripe to the tail; a black belly and light gray to black udder; black legs below the knees and hocks; ears black inside and bay outside. Bucks often have more black on the head than does, black whiskers, and black hair along the shoulder and lower chest with a mantle of black along the back. Bucks frequently have more white hairs through the coat than does.

The face is straight. A Roman nose is discriminated against.

Saanen

http://nationalsaanenbreeders.com

The Saanen dairy goat originated in Switzerland. It is medium to large in size with rugged bone and plenty of vigor. Does should be feminine, however, and not coarse. Saanen are white or light cream in color, with white preferred. Spots on the skin are not discriminated against. Small spots of color on the hair are allowable, but not desirable. The hair should be short and fine, although a fringe over the spine and thighs is often present. Ears should be erect and alertly carried, preferably pointing forward.

The face should be straight or dished. A tendency toward a

Roman nose is discriminated against.

Sable

http://sabledairygoats.com

According to the American Dairy Goat Association's Breed Standards, the Sable dairy goat is medium to large in size with rugged bone and plenty of vigor. Does should be feminine, however, and not coarse. Their hair is short; ears should be erect and alertly carried, preferably pointing forward. The face should be straight or dished. The Sable may be any color or combination of colors, solid or patterned, EXCEPT solid white or solid light cream.

The American Goat Society's Sable Breed Standard is a bit more specific. It states that the Sable is a color variation of the Saanen breed. Sables can be the offspring of Sables or Saanens. Other than color, this breed is identical to the Saanen. Sables may be any color except solid pale cream or white.

Mature Sable does should be at least 30" tall at the withers, and should weigh at least 135 pounds. Mature Sable bucks should be at least 32" tall at the withers, and should weigh at least 160 pounds.

Toggenburg

http://nationaltoggclub.org

The Toggenburg is a Swiss dairy goat from the Toggenburg Valley of Switzerland. This breed is medium size, sturdy, vigorous, and alert in appearance. The hair is short to long in length, soft, and fine. Its color is solid varying from light fawn to dark chocolate with no preference for any shade.

Distinct white markings are as follows: white ears with dark spot in middle; two white stripes down the face from above each eye to the muzzle; hind legs white from hocks to hooves; forelegs white from knees downward with dark vertical stripe below knee acceptable; a white triangle on each side of the tail; white spot may be present at root of wattles or in that area if no wattles are present. Varying degrees of cream markings instead of pure white acceptable, but not desirable. The ears are erect and carried forward. Facial lines may be dished or straight, never Roman.

Parts of the Goat

It is very important for the overall health and longevity of your goats that you recognize and then attempt to breed for structurally correct goats. Once you are able to recognize what makes a "good" goat, you can compare yours to the "ideal," and make judgments about how to go about improving those areas of weakness in your herd.

Even before you can understand the qualities necessary for improvement within your herd, you must recognize the important parts of a goat.

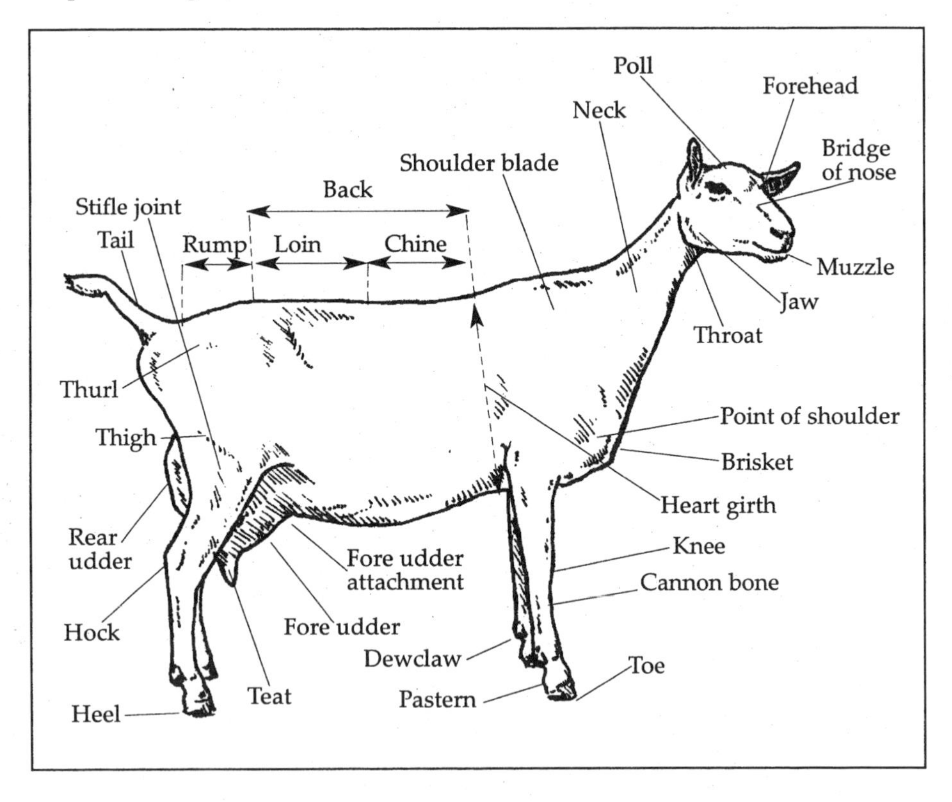

No matter whether you are raising dairy goats, Boer goats, Pygmies, or others, a heart girth is a heart girth, a chine is a chine. By becoming familiar with the parts of the goat, you can then go to the breed standards for the goats you are raising and make comparisons of your goats to the ideal.

Recognizing a "Good" Dairy Goat

The illustrations appearing in this article appeared originally in the Nigerian Dwarf Goat Association's "Official Judge's Training Manual." We wish to thank both the artist, Denice Munchrath, and the Nigerian Dwarf Goat Association for allowing us to reprint them here.

Recognizing a "Good" Dairy Goat

A "good" dairy goat is not just a doe with a pretty udder or one that milks 4,000 pounds a year. A "good dairy goat" must have a combination of positive qualities, all of which allow her to produce lots of milk, have numerous kids, and live a long productive life. Many traits go together to make a "good" goat, and if you learn to recognize these traits, you will be able to improve your breeding program and purchase better goats.

No matter what someone tells you, no one can look at a young kid and tell that she will be a permanent champion or have great udder attachment, but you can learn to recognize certain positive traits that goats of all ages should possess.

The first area to look at is **general appearance**. Structurally, the doe should have a strong, level top line; her withers should blend smoothly into the shoulder blades (no bumps or humps as you run your hand down her neck over her withers and shoulders). Her front legs should be wide apart, strong, and straight (not curved as you look at them from the side); her rear legs should be set wide apart at the hocks, with a wide arched opening in the escutcheon area. As you look at her rear legs from the side, they should be nearly perpendicular from hock to pastern. Look for short, strong pasterns, not ones that are broken and weak. Does with these positive structural traits should be productive does; they will have the strength to withstand the rigors of heavy milking and strenuous kid bearing for many years.

Dairy character is also important. A doe should look feminine; she should walk with gracefulness and animation. She should be an "open" doe—her ribs should be set wide apart; they should be flat (as should all her bones) and long. To feel the difference between flat-boned and round-boned does, run your hands down the ribs of a number of does. Flat-boned does' bones actually feel flatter; the space between ribs will usually be wider. The more times you do this, the easier finding that flat-boned doe will be. With more experience, you will actually be able to pick out "dairy" does from across the barn or ring; they ooze femininity, angularity, and, well, dairyness.

Recognizing a "Good" Dairy Goat

A "good" doe is a capacious doe, and you can see some of this potential **body capacity** even in kids. Look for a doe with deep heart girth (more room for the lungs and heart). In small kids, look also for width of the chest floor; a really narrow, pinched kid will never develop tremendous body capacity. When choosing a kid, do not worry about size of barrel as much as body length in general. In older does, look for increasing depth from front to rear as you look from the side. Remember that large body capacity means more room for food and for kids. Be careful, though, not to mistake a fat, beefy doe for a capacious doe. You are looking for a doe with body capacity and dairy character.

For a doe to milk well over a long lifetime, she will need to have a **well-attached udder**. Udders without much attachment tend to flop around, get stepped on, and generally are more prone to injury and disease than udders that have strong attachments. Look for a high, wide rear udder attachment and ideally a smooth, well-extended fore udder. A doe can have a small pocket in the fore udder, though, and still have a functional udder—if she has strong rear udder attachment and a correctly attached medial suspensory ligament. The smooth fore udder is icing on the cake. The medial suspensory ligament is the udder's primary support; if it is weak, the whole udder will sag. Finally, the udder must be capacious (that means large in relation to the doe's size), and when the doe is milked out, ideally there should not be a whole lot of "beef" or "meat" in the udder. The more there is, the less capacity there is for milk.

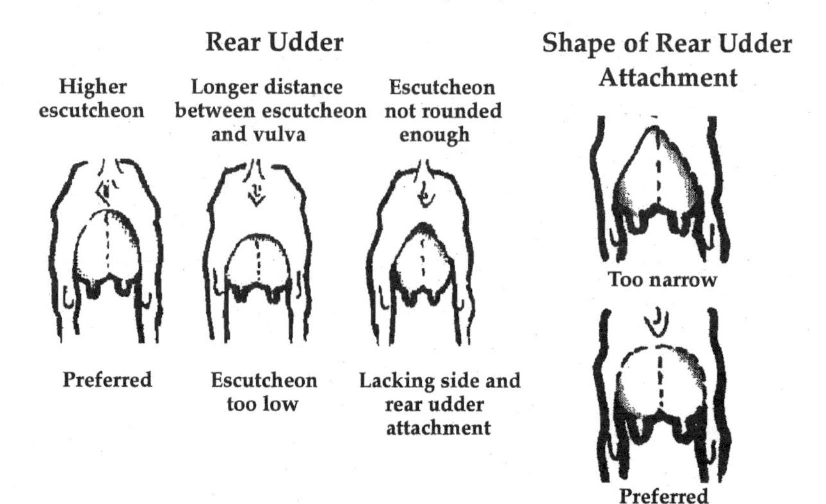

Fore Udder

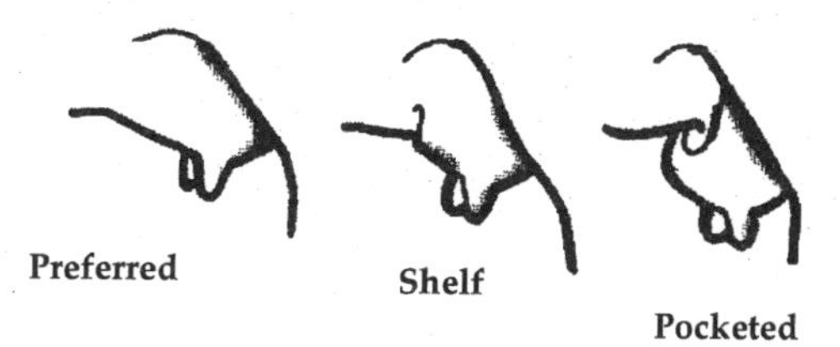

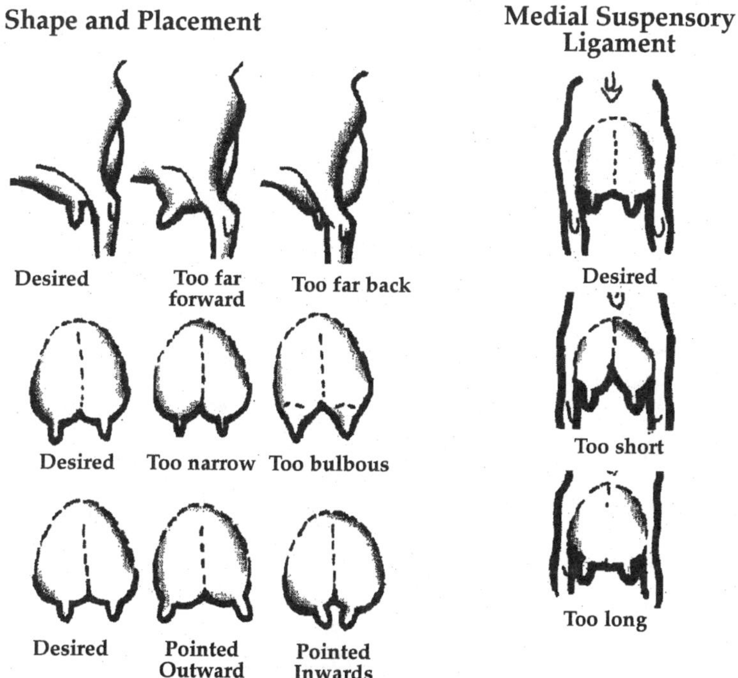

Why do you need to know what a "good goat" looks like? Remember, it costs the same to feed a structurally sound goat as an unsound one, and a "good doe" will give you many more years of service, more milk, and more kids, with fewer health problems. However, no matter how structurally sound a goat is, if she does not have good management, she will never reach her potential. So you must give your "good does" a sound program of health care, feeding, and general maintenance, to insure that they live up to their potential.

Stature Chart of Goats

The following Stature Chart originally appeared in educational materials published by the Nigerian Dwarf Goat Association. We wish to thank both the artist, Denice Munchrath, and the Nigerian Dwarf Goat Association for allowing us to reprint it here.

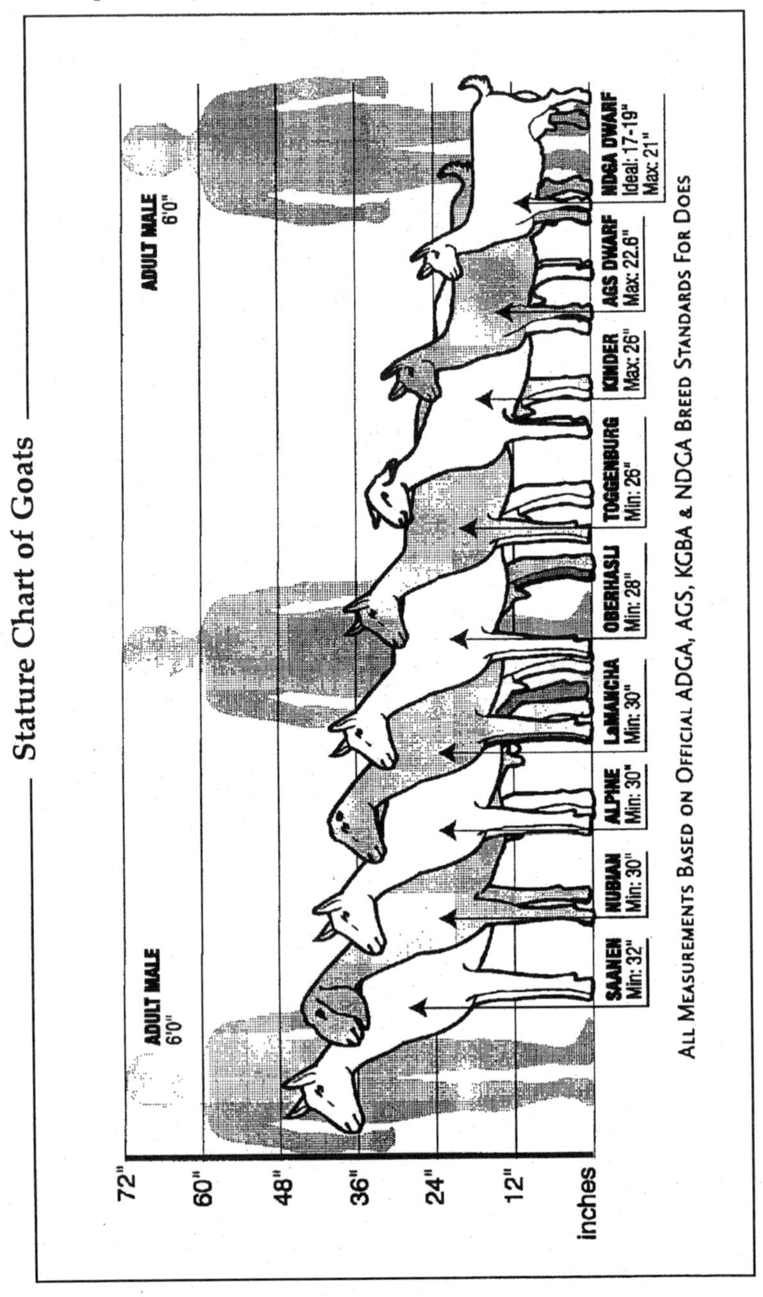

This article originally appeared in United Caprine News, ***October, 1998.*** *We wish to thank Jeff Klein, its publisher, for allowing us to reprint an edited version here. The illustration comparing the Pygmy to the Nigerian Dwarf is the work of Denice Munchrath, and we thank the Nigerian Dwarf Goat Association for allowing us to reprint it here. [Editor's note: Nigerian Dwarf goats are now a breed recognized and registered by ADGA.]*

The "Small" Dairy Goat

by Pat King

Most of us can readily identify the six ADGA-recognized breeds of dairy goats. While these breeds ideally share desirable traits of conformation, each breed has distinguishing traits—elements of breed character—that set them apart one from another. These dairy goats are "roughly" the same size and have been selectively bred to serve the same function—efficient milk production—but in ways both apparent and subtle, they are undeniably different.

However, tell someone you breed Nigerian Dwarf dairy goats and too often the response will be, "Oh, you have Pygmies? They're soooo cute." Despite the facts that more Nigerian Dwarves are registered by the American Goat Society than Oberhaslis by the American Dairy Goat Association, that increasing numbers of breeders are advertising in national dairy goat publications, and more Nigerian shows are being held in conjunction with ADGA-sanctioned shows, not enough people yet realize that Nigerian Dwarves and Pygmies are two very different breeds selected for very different forms and functions.

The History of Dwarves

Dwarf goats are found through central and western Africa with much mingling of varieties. Generally standing less than two feet tall, these goats exhibit body types ranging from compact and blocky to upstanding and refined. All possess the rear-angled ribbing associated with dairy goats.

Importations of dwarves began sometime early in the century and continued through the 1960's. Coming from various West African locales, they probably represented a broad range of body types and coloration. Small size was the main selection criterion, but many of these animals carried genes for greater stature.

Genetically, the imported population evidenced two different types of dwarfism. Some exemplified achondroplastic dwarfism, characterized by a large head, wide body, and shortened limbs. Others exemplified pituitary dwarfism, with growth arrested in all body parts and systems more or less equally. The latter were true miniatures with well-proportioned bodies. These two types developed in America into two different breeds: the African Pygmy, with its achondroplastic traits, and the Nigerian Dwarf, a true miniature dairy goat. And although their names reflect their African roots, they are both breeds made in America.

The Nigerian is Not a Pygmy

The Pygmy is cobby and compact, with width and depth of body greater in relation to its height. Its legs are short, particularly its cannon bones. Its head is large but short. The bone is heavy, and the Pygmy carries much heavier muscling than a dairy goat. Pygmies have round, flat shoulders and a short rump.

Selection in the Pygmy breed has focused on small size and personality. The common coloration is agouti pattern, a grizzled color produced by the barring of light and dark bands on each shaft of hair, with black on the head, body and legs.

The Nigerian Dwarf shares the desirable characteristics of dairy conformation evidenced in the standard-sized breeds, though they are proportionately smaller in scale. Compared to the Pygmy, Nigerians are more angular and refined, with flatter, flintier bone and much less muscling. {Their] legs are refined and longer in relation to body size. They evidence more "stretch" from a longer, leaner neck to a longer more level rump. Because of its inherently functional type, it is a reproductively sound breed. All colors and color patterns occur.

Pygmy **Nigerian Dwarf**

Nigerian Dwarf Goat Associations

The Role for the Nigerian Dwarf

The Nigerian Dwarf has the ability to fill a unique niche in the dairy goat world. It is ideal for those who have limited space, who do not need large quantities of milk or lack acceptable outlets for the surplus, or who require animals easily managed by one person.

The Nigerian Dwarf is a fully-functional, productive dairy goat that just happens to be small.

American Goat Society (AGS)

In 1993, the American Goat Society adopted a breed standard and opened a herdbook for the Nigerian Dwarf. This standard was revised in 1996.

AGS Nigerian Breed Standard

The Nigerian Dwarf is a miniature dairy goat originating from West Africa and developed in the United States. The balanced proportions of the Nigerian Dwarf give it an appearance similar to the larger, Swiss breeds of dairy goats. Shorter height is the primary breed characteristic of the Nigerian Dwarf, with does measuring no more than 22 1/2″ at the withers and bucks measuring no more than 23 1/2" at the withers.

They are known for their high quality milk, often with exceptionally high butterfat content. Nigerian Dwarves are gregarious, friendly, hardy animals that thrive in almost any climate.The medium length ears are erect and alert. The face is either straight or slightly dished. The coat is of medium length, and straight. The Nigerian Dwarf is the only dairy breed known to occasionally have blue eyes. Both brown & blue eyed animals are encountered with no preference being given to either eye color. Any pattern, color, or combination of colors is acceptable.

Mature Nigerian Dwarf does should be no more than 22 1/2″ tall at the withers. Mature Nigerian Dwarf bucks should be no more than 23 1/2" tall at the withers.

American Dairy Goat Association (ADGA)

In 2005, the American Dairy Goat Association began registering Nigerian Dwarf goats. In 2008, ADGA registered a total of 2,896 Purebred Nigerian Dwarf goats.

ADGA Nigerian Dwarf Breed Standard

The Nigerian Dwarf is a miniature breed of dairy goat originating in West Africa and developed in the United States. The balanced proportions of the Nigerian Dwarf give it the appearance of the larger breeds of dairy goats, but does stand no more than 22.5" (57cm) and bucks no more than 23.5" (60cm). Any color or combination of colors is acceptable. The medium length ears are erect and alert. The face is either straight or dished, and the hair is short and fine.

.Nigerian Dwarf Goat Association (NDGA)

Today, the Nigerian Dwarf Goat Association registers Nigerian dwarfs, sanctions shows, trains judges, provides extensive educational materials, and generally promotes the breed.

NDGA Breed Standards

The Nigerian Dwarf is the only true miniature goat breed of dairy type and character. Its conformation is similar to that of the larger breeds, but the parts of the body are in balanced proportion relative to their size. The profile of the face is straight although some have a small break or stop at eye level. The ears are alert and upright. The coat is straight with short to medium length hair; with short, sleek and smooth hair being ideal. The desired height for does is 17 to 19 inches with a maximum of 21 inches. The desired height for bucks is 19 to 20 inches with a maximum of 23 inches. There is no minimum height for does or bucks. Any color or combination of colors is acceptable, with Pygmy goat breed specific markings being a moderate fault.

American Nigerian Dwarf Dairy Association (ANDDA)

The American Nigerian Dwarf Dairy Association promotes the breeding, improvement, and management of the Nigerian Dwarf first and foremost as a dairy animal. ANDDA is a breed club, not a registry, committed to keeping Nigerians a healthy and productive breed, supporting the American Goat Society and American Dairy Goat Association. Fundamental to its purpose is the belief that dairy goats are best served by multi-breed function-oriented registries.

Nigerian Dwarf Milk Production

In time, the USDA's Animal Improvement Performance Laboratory amassed enough pedigree, production, and conformation data to include the Nigerian Dwarf on its list of recognized dairy goat breeds.

The minimum requirement for Nigerian Dwarf milk production

is 1/3 that of the standard sized breeds. It increases 2 pounds of milk for each additional month of age at time of freshening. The butterfat requirement is based on 5% of the minimum pounds of milk for the respective age.

The standard-sized breeds have had the advantage of many decades of Dairy Herd Improvement testing in order to accumulate official production data and thereby increase the accuracy of genetic selection for production traits. Nigerians have only been production-tested for a relatively short time, and their full dairy potential is just beginning to be realized. A number of Nigerian does, however, produce enough pounds of butterfat in a lactation to meet AR requirements for butterfat set for the large breeds.

Breed Production Averages for Nigerian Dwarf Does with 275-305 Days in Milk on DHI Test*

ADGA/AGS	#	Avg. Age Start of Current Lactation	Milk lbs.	Range	BF %/lbs	Protein %/lbs
ADGA 2007	59	3y5m	806	300-1720	6.6/53	4.3/34
AGS 2007	53	2y6m	747	300-1719	6.4/48	4.4/32
ADGA 2006	43	2y6m	748	300-1720	6.5/47	3.9/29
AGS 2006	60	2y6m	787	346-1389	6.6/51	4.3/34
ADGA 2005	27	3y6m	881	281-1580	6.5/57	4.0/35
AGS 2005	31	2y10m	890	298-1575	6.5/57	4.4/38

*From the ANDDA website, July, 2009. Website notes that many more does were tested but because they did not test through 275-300 days they are not included here.

Pygmy Goats

We wish to thank the National Pygmy Goat Association for allowing us to include the following information on Pygmy goats. Much of it was taken from the Association's "A Pygmy Goat Short Course" brochure and their website. *For more information about Pygmy goats, contact the National Pygmy Goat Association directly or visit their website at http://www.npga-pygmy.com.*

Pygmy Goats

http://www.npga-pygmy.com

Not all little goats are Pygmy Goats! Pygmy Goats are a specific breed derived from the Cameroon's or West African Dwarf Goat. The first Pygmy Goats were brought to the United States in 1959 for use in petting zoos.

The Pygmy Goat is hardy, alert and animated, good-natured and gregarious, a docile, responsive pet, a cooperative provider of milk, and an ecologically effective browser. Pygmy Goats are assets in a wide variety of settings, and can adapt to virtually all climates. They are ideal personal livestock suited to today's smaller homestead.

Pygmy Goats are much more than interesting pets or pasture ornaments. They are well suited to many uses. They are loving pets, great show animals, and ideal first livestock for a young 4-H or FFA member. They are welcome as therapy animals in nursing homes and other facilities; they are often companion animals to other livestock. They give rich milk and have a smaller easy to handle carcass for the home freezer. They pull carts, or go hiking with their owner, even carrying some of the gear. A small herd can clear a woodlot in an ecologically safe and environmentally sound way.

Pygmy Goat Breed Standard

The National Pygmy Goat Association's Breed Standard describes the Pygmy Goat as being genetically small, cobby, and compact. Full-barreled and well muscled, the body circumference in relation to height and weight is proportionately greater than that of other breeds. Mature animals measure between 16 and 23 inches at the withers. Head and legs are short relative to body length. Genetic hornlessness is considered a disqualifying fault. However, disbudded goats are acceptable.

Agouti Pygmies range from light silver to nearly black in a predominately grizzled pattern. Muzzle, forehead, eyes and ears are accented in white. Front and rear hoofs and cannons (socks) are dark, as are the crown and dorsal stripe. Caramel Pygmy Goats are white through dark brown; muzzle, forehead, eyes and ears are accented in white. Front and rear stockings are dark with a vertical lighter stripe on the front, the crown and dorsal stripe and martingale are dark. Random markings are acceptable in limited amounts and characteristic locations. Coat length and density vary with climate, making Pygmy Goats at home in the desert or the northern tundra.

Breeding Pygmy Goats

Pygmy Goats are precocious breeders, bearing one to four young every nine to twelve months after a five-month gestation period. Does are usually bred for the first time at twelve to eighteen months, although they may conceive as early as three months if care is not taken to separate them early from bucklings. Bucklings can be fertile as early as twelve weeks old and certainly by sixteen weeks. Newborn kids will nurse almost immediately, begin eating grain and roughage within a week, and can be weaned at three months of age.

Kinder Goats

We wish to thank the Kinder Goat Breeders Association for allowing us to include the following information about the breed. For more information, contact the Kinder Goat Breeders Association or visit their website, www.kindergoats.com.

Kinder Goats

http://www.kindergoats.com

"Making the World a Little Kinder" by Patricia Showalter

In the late summer of 1985 Zederkamm Farm found itself with a problem. The old Nubian buck who had kept our two Nubian does fresh for a continuing milk supply unexpectedly died in his sleep one morning. The idea of hauling our does off to another farm to be bred just didn't appeal to us after an unsuccessful preliminary search for a replacement buck. Our real interest was in the milk. What would it matter if the resulting kids were purebred or not? One of our Pygmy bucks stepped up (way up) to volunteer his services, and so began the Kinder goat.

Briar Rose was born first, then Liberty and Tia in the summer of 1986. We were surprised and delighted with the appearance and rate of growth of these little does. Liberty stayed with us, while the other two little girls went to nearby owners. In 1987 Liberty freshened for the first time with triplets, and proved to be a steady and reliable producer of the best milk we had ever tasted. On her next five freshenings she produced two sets of quintuplets, a set of sextuplets, a triplet, and a twin set. She led the way as the first Kinder doe entered on official milk test (DHIA). Liberty earned her star by fulfilling the same requirements as those set by ADGA for standard dairy goats.

Other local goat enthusiasts soon became involved in the Kinder project. Three of them organized what became the Kinder Goat Breeders Association (KGBA) in 1988. Kinders were introduced nationally through a front page article in *United Caprine News,*

January 1989. This small nucleus of a few goats and a handful of breeders in the Snohomish, Washington area has grown to about 150 members. There are now over 1,000 goats throughout the US and Canada in the registry.

Frequently Asked Questions About Kinder Goats:

Why the original Kinder starter kit (Nubian/Pygmy)?

At first that was what was available. Later we considered the possibility of other combinations, but always returned to the original cross. The resulting goat was attractive and distinctive, proving to be useful for both milk and meat production. Besides, it has a personality and sweet disposition surpassing anything we have ever experienced.

What does the Pygmy add to the cross?

1) Smaller size- easier to house and handle; 2) Excellent feed conversion: high production with minimum intake; 3) Exceptional quality high protein/high butterfat milk; 4) Desirable heavier muscling- translating to meatier carcass; 5) Straight or dished face for distinction as a breed characteristic; 6) Hardiness: Kinders do fine in extremes from hot to cold; 7) Multiple births: triplets and quads are routine; 8) Year around breeding season.

What contributions are made by the Nubian?

1) Some length of leg, so the milk bucket fits underneath; 2) Larger teats, for ease of milking; 3) Larger quantity of milk, also high quality; 4) Stronger udder attachments; 5) Extended lactations; 6) More correct rump and rear legs; 7) And last, that appealing length and (usually) some drop to the ears, plus a dash of fun color.

How does the Pygmy manage to breed a big, tall Nubian?

With determination and creative planning. A slope, a step, a log, a platform, any of these approaches and more have been successfully used by the Pygmy buck to accomplish his goal. He is not likely to take "no" for an answer. "Can't" is not in his vocabulary.

How do Kinders kid?

With the greatest of ease. There have been few kidding problems reported to the Association. It is always wise to be present to observe any kidding, but Kinders seldom need assistance even though multiple births are common. Kinder does are usually excellent mothers.

Boer Goats: An Overview

We wish to thank Gail Bowman, Northwest Coordinator, International Boer Goat Association, for providing the following information.

Boer Goats: An Overview

Boer goats are large-framed animals resembling, in many ways, the Nubian goat. The most obvious difference is the size. Large, double muscled animals, Boers are specifically meat goats. They can consistently produce more muscling in less time than their dairy cousins and will pass this capability on to their kids. Boers are to the meat goat industry what imported cattle were to the beef industry.

Breed Characteristics

These large animals are generally white with a reddish brown head and usually a white blaze down the middle of the face. Solid red Boer goats are also becoming more popular. Mature weights between 200 and 350 pounds for males and 120 to 200 pounds for females are considered normal. Boers have long ears that should hang down along the sides of their faces (hence their Nubian resemblance). Their leg bones and general bone structure are bigger and thicker than those of other kinds of goats. When you look at a Boer goat, you should see a deep, broad chest, good back, strong shoulders and heavy muscling in the rump.

Boers are hardy, adaptable, and easy to handle. Given adequate shelter, they are still content to lie out in the sun on 90 to 100 degree days (their skin is darkly pigmented under the white hair to reduce the risk of sunburn) or sleep outside in 10 degree weather. Boers also do not seem as interested in jumping fences as dairy goats do.

Why Raise Boer Goats Instead of Cattle?

One reason is efficiency of feed and space. Let's do the math with the following example.

The accepted standard in the Northwest for raising cattle is one cow-calf pair per acre of good pasture. For goats we would have six does with two kids each per acre. Ten months after breeding, a cow will be nursing a 75 pound calf. Ten months after breeding to a Boer buck, six dairy or Spanish goats will have raised twelve kids, and these kids will have been sold.

Boer-cross kids reach a market weight of 50 to 90 pounds at about five months of age. (A weight gain of .5 pounds a day in crossbred kids is not unusual.) If we use a market weight of 60 pounds, at $1 per pound, these twelve kids will have brought in $720. The same six does will have been rebred 60 to 90 days after the kids were born and will be 60 days pregnant ten months after the first breeding!

At the end of 18 months, the cow should be pregnant again, and her calf will be ready to be sold for $325 (current market conditions here in the Northwest). The six does will have kidded again with another twelve kids who are now 5 months old and ready to be sold for another $720 (plus the does will be pregnant again).

Feed cost comparisons between one cow and six goats will vary greatly, depending on the time of year, type of pasture, and area of the country. It may cost a little more to feed the six does than the single cow, but the difference in the offsprings' sale prices ($1440 vs. $325), a whopping $1115 in gross sale price, certainly makes up for the additional feed costs.

If you have ten acres, you can easily raise sixty goats or ten head of cattle. Goats, especially Boers or Boer crosses, can survive, even prosper, on poor pasture and brush that would not support cattle. Many breeders find the fact that goats will eat berry bushes, Russian olive, elm or cottonwood trees, ragwort, gorse, dock, amerauthis and other weeds, to be an important factor when deciding to raise goats. Some ranchers also find it good pasture management to run goats on the pasture after their cows to clean up the weeds.

Characteristics of Goat Meat

Goat meat has a lower fat content than either lamb or beef and is eaten by over 80% of the world's population. We are beginning to see a dramatic rise in the popularity of goat meat in the United Sates. One reason might be the improved flavor that the Boer adds to the meat. Adding just 50% Boer bloodlines to our goat meat produces a very mild and tender, light red meat that readily takes on any seasoning.

Starting Your Boer Goat Herd

Just as in other breeds, with a small herd you will need one or two Boer or Boer-cross bucks, and as many "brood" does as you wish to keep. These does may be from dairy, hair, Spanish, or other stock. The meatier they are to begin with, the better product their offspring will be. Try to purchase your Boer breeding animals from a reputable breeder rather than from an auction house or stockyard. You will usually end up with healthier animals doing this. Buying goats with registration papers, while not necessary, will give you better resale value, and allow you to see, on paper, the background on your stock.

Finally, as you breed your animals, keep good records so you will be able to track which breedings work and which do not. Only with this information can you make informed judgments in the future and improve the overall quality of your meat herd.

Meat Goat Production

The following article appears on the Alabama Cooperative Extension System website: http://www.aces.edu/pubs/docs/U/UNP-0104. We wish to thank the author, Robert Spencer, Urban Regional Extension Specialist, Urban Affairs and New Nontraditional Programs Unit and the Urban Centers in North Alabama for allowing us to reprint an edited version of this article. For further information, he suggests going to

http://www.aces.edu/pubs/docs/indexes/unpas.php#small.

"Overview of the United States Meat Goat Industry"

Goats are one of the oldest domesticated livestock. There are numerous breeds of goats that are categorized by their capacity to produce fiber, milk, or meat. While those breeds specializing in fiber and milk production may also serve as meat goats, most meat goats are not ideal producers of fiber or milk.

Boer, Kiko, Myotonic (Tennessee Fainting Goat), Savannah, Spanish, or any of these breed combinations are ideal meat producers. While various forms of goat production have existed throughout the world for centuries, in 1992, the United States developed a strong interest in meat goat production. Since then, the meat goat industry has been the fastest growing segment of livestock production in America. Initial interest in meat goat production primarily took place in the Southeast, with Texas and Tennessee leading most states and having the largest goat populations. In more recent years, interest in meat goat production has expanded across the country from California to Maine.

Since the early 1990s, importation of goat meat and domestic meat goat production has continued to increase. Much of this is attributed to increasing demand created by populations immigrating to the United States. Persons from the Middle East, Southeast Asia, and the Caribbean, who have relocated to the United States, are accustomed to eating goat meat. These persons continue to have a preference for goat meat during religious holidays, on special occasions, and during extended holidays when families and friends gather together. In 2003, the US Census Bureau reported 33.5 million foreign-born US citizens. This population trend that accounts for domestic goat meat consumption is expected to last for years to come and is considered to be the driving force behind the increasing interest in meat goat production throughout the United States.

Meat Goat Production

Information from the United States Department of Agriculture's (USDA) Foreign Agricultural Service (FAS) shows that Australia is the primary exporter of goat meat into the United States. In 2008, the USDA's National Agriculture Statistics Service reported 3,150,000 meat goats in the United States, **yet American producers are unable to meet domestic demand** [editor's emphasis]. Prior to the early 1990s, there were no organized efforts to promote meat goats and their production. The formation of several producer organizations took place as interest and the potential for meat goat production began to develop. On a national basis, they include, but are not limited to, the American Boer Goat Association, the American Meat Goat Association, the International Boer Goat Association, and the US Boer Goat Association. Such groups recognize the interest and potential for meat goat production and continue to support the industry as well as educational and outreach efforts designed to benefit potential and novice producers. These same organizations have also been instrumental in promoting goat demonstrations for all ages to highlight quality animals and desirable breed traits and characteristics.

Meat goat production is an alternative form of livestock production for novice producers seeking learning opportunities to expand their knowledge base. While producers tend to focus on marketing and health care as their primary areas of concern, they benefit significantly by diversifying their knowledge base into other aspects of meat goat production. Primary areas requiring continuing education include breed options for meat goat production, reproduction management, feeding and nutrition, health concerns and management, pasture management, fencing and facilities, marketing strategies and options, and familiarity with live animal evaluation. Educational and outreach programs using a comprehensive approach to address all aspects of meat goat production tend to best serve novice meat goat producers. Producers in general benefit from attending educational workshops, reading industry relevant publications, and interacting with "seasoned" producers. Informed producers are less likely to repeat the same mistakes and more likely to be satisfied with their management decisions. With the adaptation of proper management strategies and implementation of beneficial practices, meat goat production has potential as a sustainable form of livestock production.

While the meat goat industry offers much promise, factors that compromise potential expansion and production are:

• Inconsistent supply to satisfy peak season demand, generally during religious holidays.

• Lack of producer knowledge and strategic planning to arrange for reproduction based upon anticipation of these peak demand opportunities. Many producers are unfamiliar with religious holidays associated with persons from the Middle East, Southeast Asia, and the Caribbean.

• Non-standardization of production practices. The industry is still in its developmental stages, and organizations and educators have insufficiently developed cohesive and standardized best management practices.

• Predation and parasitism are two factors negatively affecting herd health, in some cases resulting in mortality and production and profitability losses.

• Limited information regarding quality assurance practices.

• Failure to establish food quality and safety guidelines and the standardization of processing and cuts.

• Limited number of USDA-inspected facilities across the country.

The future of the meat goat industry cannot continue to rely on demand from US citizens originally from the Middle East, Southeast Asia, and the Caribbean. Price and availability of goat meat, as well as competition from more commonly found commercial meats at grocery stores influence the demand for goat meat. Also, the next generation of immigrants will more likely become more "Americanized" than their parents or grandparents and seek out traditional American meats such as beef, chicken, and pork.

Traditional Americans are unaware of the health aspects associated with goat meat and might be more receptive to alternative forms of meat if they were more knowledgeable. Therefore, it is inevitable that those with an interest in the success of the meat goat industry think toward the future and consider potential opportunities to promote the health aspects of goat meat. Not only is it a viable alternative to traditional meats, but it has health benefits that surpass beef, pork, poultry, and seafood. Goat meat also has culinary appeal due to its versatility and ability to be prepared in various forms, recipes, and during special occasions, while providing stimulating conversations during food-related gatherings.

The good news is that the meat goat industry will continue to grow for years to come. Projections from the US Census Bureau confirm the continued growth of populations from the Middle East, Southeast Asia, and the Caribbean in the United States. These populations will continue to be the driving force for demand for goat meat and growth in the meat goat industry. The National Agricultural Statistics Service (NASS) forecasts a continued increase in demand for goat meat and a consecutive growth in meat goat production within the United States. With this positive outlook, the challenge to meet the demand will ensure potential opportunities for those interested in meat goat production. These opportunities will necessitate those institutions, organizations, and leaders active within the meat goat industry to continue to hold a vested interest to ensure promotion and producer education as outreach efforts are continued and expanded as needs arise.

We wish to thank Terry Hankins, the publisher of Goat Rancher *magazine, for allowing us to include the following edited article. The author, Dr. Stuart Southwell, is a foremost contributor to the development of the Boer goat industry. He is widely known and respected for pioneering many of the embryo transfer techniques and protocols used by veterinarians and goat producers around the world.*

The Boer Goat Goes International

by Dr. Stuart Southwell, B.V.Sc, M.R.C.V.S.
Premier Genetics NZ Ltd, Drury, New Zealand

In 1985, Landcorp Farming N.Z. decided to improve its Angora goats by importing genetics from Southern Africa, and they decided to use Zimbabwe as the country of importation. Alan Aiken, the Landcorp representative, traveled to Zimbabwe to assess the situation and select the donor animals. While there, he was introduced to a funny-looking goat with long brown ears. The locals called this goat the "Boer." These goats had been imported from South African many years ago and now formed a significant population in Zimbabwe. The "Boers" attracted Alan Aiken's interest as they would many people in future years.

In January, 1987, Landcorp went to Zimbabwe to flush embryos from Angora and Boer females. This took place at a quarantine station called Iridor, just outside of Harare. Approximately two hundred embryos were frozen from this collection and exported to New Zealand. These embryos were implanted in goats on Soames Island (New Zealand's maximum quarantine facility) in May, 1987.

Also in May, 1987 Landcorp went back to Zimbabwe to collect more Angora and Boer embryos. The donors in this second flush were completely different from those at Iridor. They were in quarantine at Keymer Farms owned by D. Banks. Because of quarantine regulations and the seasonal nature of the goat's reproductive cycle, the embryos from the flush were not implanted into recipients in New Zealand until April, 1988.

Landcorp's final collection of Boer goat embryos from Zimbabwe took place in May, 1988. These were implanted in April, 1989, and the pregnant recipients went to a second quarantine station in the South Island called Eyrewell.

These three collections from Zimbabwe in 1987 and 1988 formed the basis of Landcorp's Boer goat industry. By the end of 1988, New

Zealand had two groups of Boer goats, one owned by a consortium with B. Moodie as its spokesperson, and the other owned by Landcorp Farming.

Also in 1988, an Australian group flushed Boer goat embryos from new genetics, also at Keymer Farms. Those animals were released from quarantine in Australia in November, 1995, and by that time they had propagated to the extent that there were approximately 2,000 Boer goats at the Terraweena quarantine facility.

Prior to the release of Landcorp's animals from quarantine, Landcorp implanted Boer embryos into recipients at Olds College in Canada. These were the first Boer embryos and later live kids to be born in North America. These goats stayed in quarantine at Olds College until the New Zealand release in April 1993, at which time they were able to be released into Canada and the United States.

From that point on, the Boer goat has truly become an international entity. Since 1987, Boer goats have moved from Africa to New Zealand, Canada, Mexico, Australia, the United States, Indonesia, England, India, France, Malaysia, Denmark, British West Indies, Netherland Antilles, and numerous other countries.

Standards for the Boer Goat

In describing those traits which constitute "standards" for the Improved Boer Goat, the American Boer Goat Association is leaning heavily on the standards which have evolved during the development of this breed over the past seventy years in South Africa.

The standards which they have developed have the explicit objectives of improving the breed for economic production. The South African Boer Goat is a recognized breed in its country of origin, and many experts throughout the world consider this to be the premier goat meat producing breed. Three selection criteria have contributed to this recognition:

1. Large frame size
2. High carcass yield grades
3. Uniform visual appearance

Visual uniformity exists not only in the color patterns which the animals carry, but also in the uniform stature and yield grades. The lack of one or more of these traits in other breeds that have been used for meat production has held back the development of the meat goat industry in the United States and abroad.

The South African Boer Goat developed by natural selection practices of breeders in Africa under the often stressful conditions of the African environment. These breeders demanded that only the best, commercially viable animals be recognized as superior. On July 4, 1959, the Boer Goat Breeders' Association of South Africa was formed. One of the first undertakings was to establish breed

standards. These standards have changed little over the years, and the changes which have been introduced have assisted in improving the breed.

Five types of Boer goat are recognized in South Africa: these include the **Ordinary Boer Goat, Long Hair Boer Goat, Polled Boer Goat, Indigenous Boer Goat,** and **Improved Boer Goat**. The Improved Boer Goat is the only line or type which the South African Boer Goat Association will register as a breeding quality animal.

Boer Goat Breed Standards

The American Boer Goat Association (ABGA) provides the following standards as a guide to owners and breeders of Improved Boer Goats in the United States. These standards describe what an Improved Boer Goat should be, citing the most desirable traits that make up the ideal individual. When evaluating an animal's value and desirability, the best balance of all the standards should be sought. In summary, the animal that possesses the greatest conformity to the breed standards, when viewed as a complete package, is the best representative of the Boer breed as it was developed to be.

The American Boer Goat Association registry has been developed to document and maintain bloodlines through pedigree only. Owners and/or breeders should use the standards to evaluate animals when establishing desirability and value. Only animals that meet or exceed these standards will be eligible to advance to performance evaluations and ultimately attain recognition in the Ennobled Book certified by the ABGA beginning September 1, 1994.

Overall Quality, Size, Appearance and Type

The overall objective is for a goat to have suitable size with maximum meat yields, good structural conformation which meets environmental and production requirements, high adaptability to environmental conditions, and high fertility.

Ideally, this is an animal with short glossy hair with fine luster. The major portion of the body should be white with dark coloration around the head and pigmented skin in hairless areas around the head and under the tail to reduce sunburn, cancers, and skin diseases. A loose, supple skin helps the animal adapt to wide climatic conditions, and possibly provides resistance to external parasites.

In general appearance, a Boer goat has a dark head and horns which curve backwards. Animals should be strong, vigorous, and symmetrical, with well balanced muscling. Bucks should be masculine and well proportioned, but not overly developed in the head, neck and forequarter making it out of proportion with the rest of the body. Does should be feminine, yet strong, and have a slightly more angular chest than bucks. They should be able to breed easily and have the conformation and constitution to easily raise fast-growing kids.

Overall, the ideal is a rapidly growing, well proportioned goat of suitable size with the ability to maximally produce prime cuts of meat to meet consumer demands. A desirable relationship between the length of leg and depth of body should be achieved at all ages, with kids and young goats being slightly longer in the leg.

Faults are cull characteristics or defects which decrease the value of the goat for breeding purposes, and will ultimately affect an animal's eligibility for ennobled status.

Conformation (as of April 11, 2007)

Any extreme occurrence of an undesirable trait is a disqualification.

Head:

A prominent, strong head with brown eyes and a gentle appearance. Nose with a gentle curve, wide nostrils, and well formed mouth with well-opposed jaws. The jaws must have no over or under bite from birth to 24 months of age. After 24 months no more than a ¼ of an inch under bite is allowed. Correct fit is preferred. Teeth should erupt in the proper sequential positions. The forehead should be prominent and form an even curve linking the nose and horns. Horns should be dark, round, strong, of moderate length, positioned well apart and have a gradual backward curve before turning outward symmetrically. Ears should be smooth of medium length and hang downward.

Faults: Concave forehead, straight horns, jaw too pointed, overshot or undershot jaws

Disqualifications: Blue eyes, ears folded lengthwise, short ears, parrot mouth or more than ¼ of an inch under bite.

Neck and Forequarters:

Neck of moderate length and in proportion with body length. Forequarters full, well-fleshed, and limbs well jointed and smoothly blended. The chest should be broad. Shoulders should be fleshy, well proportioned with the rest of the body and smoothly blended and fitted into the withers. Withers should be broad and well rounded and not sharp. Legs should be strong, well placed and in proportion with the depth of the body. Pastern joints should be strong and hooves well-formed and as dark as possible.

Faults: Neck too short or too thin: shoulders too loose, and any structural foreleg, and muscle, bone, joint, or hoof deformities or abnormalities to include but not limited to knock knees, bandy legs, hooves pointing outward or inward, splay toes, buck knees, hollow legs, straight or weak pasterns.

Body

Body should be boldly three-dimensional: long, deep and wide. Ribs should be well sprung. Loin should be well muscled, wide and long. The top line should be reasonably straight and strong and the shoulder well rounded with an abundance of muscle from shoulder through hip.

Faults: Concave or swayback; chest too narrow or shallow or flat; shoulders weakly attached; inadequate muscle through the back and loin, pinched heart girth.

Hindquarters

Rump should be broad and long with a gentle slope. Britch and thighs well muscled and rounded. Base of the tail must be centered and straight. The remainder of the tail can curve upward or to one side. Legs should be strong and the leg should have a straight axis from the hip (pin bones) through the hock, fetlock, and pastern. Hoofs should be well-formed and as dark as possible.

Faults: Weak pasterns, straight pasterns, rump too steep, sickle-hocked, cow-hocked, post legs.

Disqualifications: Wry tail

Skin and Covering

Any extreme occurrence of an undesirable trait is a disqualification. Skin loose and supple. Eyelids and other hairless areas must be pigmented. Hairless areas under the tail should be at least 75% pigmented: 100% is preferred. Short glossy hair is desirable. A limited amount of winter down or under-coat will be accepted during winter, especially in colder environments.

Faults: Hair too long or too coarse.

Disqualifications: Not enough skin pigmentation.

Reproductive Organs

Any extreme occurrence of an undesirable trait is a disqualification.

Does should have well formed udders with good attachment with the number of functional teats not to exceed two per side. A split teat with two distinctly separated teats and openings with at least 50% of the body of teat separated is permissible but teats without a split are preferred. It is most important that the udder is constructed so that the offspring are able to nurse unassisted.

Kidding or Pregnancy Does must have kidded or exhibited pregnancy by 24 months of age.

Faults : Udder and teat abnormalities or defects to include but not limited to oversized or bulbous teats, pendulous udder.

Disqualifications: Cluster teats, fishtail teats or a doe that has not kidded or exhibited signs of pregnancy by 24 months of age.

Bucks must have two large well-formed, functional, equal sized testes in a single scrotum with no more than a 2" split in the apex of the scrotum.

Disqualifications: Single Testicle. Testicles too small. Abnormal or diseased testes; excessive split in scrotum..

Coloration

The typical Boer goat is white bodied with a red head, but no preference is given to any hair coloration or color pattern.

Other Boer Goat Associations

In addition to ABGA, Boer Goat breeders may choose to register their goats with two other registries: the International Boer Goat Association, Inc., and the United States Boer Goat Association. Both Associations register Fullblood/Purebred and Percentage Boer Goats. Both Associations offer a full range of services and information including their web site, breed standards, magazines, breeder information, shows, and much more. Look for information at

http://www.usbga.org/

http://www.intlboergoat.org

Kiko Goats

We wish to thank Steve and Sylvia Tomlinson of Caston Creek Ranch for submitting the following information on Kiko goats. For more information, contact the American Kiko Goat Association.

What is a Kiko Goat?
Sylvia Tomlinson corroborated by
Graham Culliford, Goatex Group Ltd.,
New Zealand

The Kiko goat is not just a goat with an unusual name. It is a goat with an unusual ability to survive and gain weight. The word "Kiko" comes from the indigenous population of New Zealand and means "meat."

New Zealand was discovered in 1769 and colonized by Europeans. Part of this process included the importation of "milch" goats. Over time goats escaped into the wild. Since there were no natural predators for the goat in New Zealand, the feral goat herd thrived. Since they still had to survive kidding, weather, parasites and disease, over the course of years, natural selection resulted in an extremely vigorous and hardy goat.

The New Zealand farmer is an astute observer, and the hardiness of the feral goat was not lost upon him. In the 1970's a group of ranchers organized in order to develop a new meat goat. Today, this group is known as Goatex Group Limited, of Christchurch, New Zealand—the developers and breeders of Kiko goats.

To develop the breed, an unselected population of several thousand feral goats were captured and subjected to stringent selection for specified traits. Does were bred to domestic purebred bucks to introduce superior genetics. The goal was to increase the frame size, muscling and milk. A very strict breeding program was followed. Geneticists in the Ministry of Agriculture and Fisheries assisted in drawing up guidelines and offering advice for the program.

Overall the two main criteria for selection were survivability and weight gain. No shelter was provided. No assistance was given during kidding. No supplemental feeding was offered in range conditions generally considered demanding. No hoofs were trimmed. Minimal parasite control was administered.

The animals that performed exceptionally were retained, and in 1986 the herd was closed to outside bucks. Outside does were slowly added if they met the selection criteria.

Today the Kiko goat is a large framed, early maturing goat that demonstrates exceptional conversion rates. They are not as heavily boned as some breeds; consequently, they exhibit high cutability. This trait, combined with their lean, succulent carcass, yields a product that puts money back into the producer's pocket.

In circumstances that would stress the best of the four-legged critters, the Kiko prevails. The advantage to the commercial rancher is obvious. Unlike the hobby rancher who may have more time and resources than goats, the commercial rancher does not. Pet names tend to go out the window, and the burden of survival rests largely on the animal. It is not that ranchers get hard-hearted. They are still willing to get up in the middle of the night and often put care of their charges before their own well-being, but physical limitations draw the line.

The hybrid vigor resulting from crossing any breed of goat with the Kiko is astounding. In 1994, Goatex Limited performance-tested a group of Kiko/Boer cross males. The most startling aspect was that the 90-day weaning weight for the crossbred males was 1.86 kg greater than for purebred Boers and 1.71 kg higher than for purebred Kikos. The experiment was repeated in 1995 and the results were again confirmed.

In cutting trials, the Kiko-cross carcass yielded more meat per carcass than a purebred Boer but slightly less than a purebred Kiko. For the commercial meat goat rancher this appears to be the premier cross.

Kikos come in every color. However most Kikos in North America are white because mostly white goats were imported, and the white coat color is dominant.

Interestingly, some breeders in New Zealand and South Africa believe that growth rates in goats are related to coat color. There is some anecdotal evidence to support this belief. As the Kiko Breeding Up Program expands, more colored genes will be introduced into the gene pool with the resulting consequence of more colored Kikos.

Some people believe this will be an asset to those ranchers living in areas with heavy predator infestation.

In order to protect the integrity of the breed, Goatex Group Ltd. formed a New Zealand Kiko Registry. All purebred animals leaving the country were fitted with a microchip and safeguards were built into the recordkeeping system to track every Kiko leaving New Zealand.

A limited number of registered purebreds have been brought into the United States. The American Kiko Goat Association is the American liaison with the New Zealand Registry for registering percentage and purebred Kiko Goats. A percentage breed-up program is available to new breeders with the 15/16 female (4th cross) considered purebred.

This article originally appeared in the AKGA Update, *the quarterly newsletter of the American Kiko Goat Association. It was submitted to us by Steve and Sylvia Tomlinson, Caston Creek Ranch, and we thank them very much for allowing us to use it. The author, Graham Culliford, is the principal of Goatex Group LLC and of Tasman Livestock Resources Ltd. A native of New Zealand, Culliford travels extensively promoting purpose-bred meat goats and demonstrating enhanced production systems to ranchers while emphasizing cost-efficient production systems through livestock selection.*

The Genemaster Story

by Graham Culliford, Christchurch, New Zealand

There is increasing interest in the purpose-bred Texas Genemaster as the meat goat industry in the United States starts to prosper. This article looks at the origins of the Genemaster, its development, and the characteristics which make it an animal particularly suited to North America.

The initial rationale for the creation of the Genemaster goat was that the Boer goat under commercial New Zealand conditions failed to thrive in the manner that would enable it to become a commercial producer of goat meat. It rapidly became apparent that the same situation was going to be faced in the United States, particularly Texas where pen feeding of purebreds was the norm and where there were insufficient numbers of commercial goats bred for slaughter to comprise a base herd.

Knowing that there were substantial numbers of Boer females in the United States which would benefit from crossing with enhanced Kiko males, Goatex Group LLC embarked on an intensive crossbreeding exercise in New Zealand to determine the degree to which Kiko influence could increase the hardiness of crossbred offspring without impacting on their meat producing ability.

In fairly short order they determined that a cross of three-eighths Kiko to five-eighths Boer would result in a terminal meat producing animal exhibiting the vigor of the Kiko (and maintaining much of the elevated cutting percentages of that breed) while retaining the carcass conformation of the Boer.

This animal was dubbed the Texas Genemaster, corporate branding was developed and registered, and today there are numerous commercial breeders in the United States who recognize the worth of the hybrid vigor that the Genemaster exhibits.

In addition, the Texas Genemaster has the low management inputs and high meat yield of the Kiko. This is accomplished at a rate of growth that is generally higher than that of the Boer or the Kiko.

In New Zealand trials, male Boer goats fed on pasture attained an average weight of 51.3 kgs at 365 days of age. Kiko males, drawn from a much smaller population but run in similar conditions attained an average of 52.4 kgs at the same age. The difference between the weights was not viewed as significant.

However, in a trial breeding program Kiko/Boer males attained the same weight at an average 296 days, over two months earlier.

Subsequent trials with varying degrees of blood in the hybrid mix have reinforced the early maturation and the enhanced daily weight gain of the hybrid offspring. This has been particularly apparent in the weight gains birth to weaning where the weaning weights for the Kiko/Boer crosses have been consistently higher than for purebred Boers and purebred Kikos.

But enhanced weight gains alone do not tell the full story. Cutting trials have demonstrated that the trial crossbreds on average yielded more meat per carcass than a purebred Boer of similar weight but slightly less than a similarly weighted purebred Kiko. The carcasses tend to have the lighter bone structure and leaner fat configuration of the Kiko while retaining the heavier muscling of the Boer.

The hybrid variants display considerably greater vigor than the Boers in browsing ability, mating enthusiasm, birthing, bonding and rearing. Consequently, there are much lower rates of post parturition mortality than found in purebred Boers.

The program to breed true Texas Genemasters requires that the rancher start his upgrading program with a purebred Boer and a purebred Kiko. These are mated to produce a 50/50 crossbreed. This goat is registrable in the Kiko Goat Registry/Genemaster Division.*

The next step is to mate either a female from step one to a purebred Boer male or a male from step one to a purebred Boer female. The resulting offspring are three-quarters Boer/one-quarter Kiko.

This animal is then mated to a straight Boer/Kiko cross. The resulting offspring are three eighths Kiko/five eighths Boer. This is Texas Genemaster which can be registered in a special division of the Kiko Goat Registry and is entitled to carry the special Genemaster brand.

Registered Texas Genemasters may be bred to each other and their offspring can be registered as certificated Texas Genemasters. However, the best use of the Genemaster is as a true terminal sire for the commercial production of meat.

Texas Genemaster sires, when compared to goats of doubtful origin, have been shown in trials conducted by Goatex Group LLC to have a substantial effect in enhancing rates of growth and cutting percentages.

Genemaster sires have more vigor than Boer sires in commercial situations. And in situations where a rancher does not want to choose one breed to the exclusion of the other, Genemaster sires can infuse commercial herds with both Boer and Kiko influence at the same time.

Daily, the realization of goat meat market opportunities grows among meat goat breeders and goat meat producers. The development of the Texas Genemaster signals the production of a goat, purpose-bred for range conditions, which will fulfill the developing market requirements of a gradable carcass which has been grown out in minimum time for minimum cost and maximum profit.

*Genemaster goats may also be registered at Pedigree International. which, according to their website, Pedigree International is here "to serve you, the breeder of domestic livestock and rare or exotic animal breeds as well as composite breeds of sheep, goats etc. We are the original registry for Savanna Goats (Savannah Goats) and Cashmere Goat. The sole registry for Spanish, TMG, TexMaster and Sako Goats. Our current Herdbooks also include: Kiko Goats, Gene-Master Meat Goats, Myotonic Goats and Dopcroix Sheep. We also maintain a Herdbook for Pet and Pack type goats that require registration to meet USDA Scrapie requirements but do not fit into any other specific breed registry." You can find more information at http://www.pedigreeinternational.com.

"Fainting" Goats

We wish to thank Lisa Johnson, Coyote Creek Ranch, Gainesville, Florida, for allowing us to excerpt the information about "fainting" goats from herd promotional material. In addition, some of this information originally appeared in The Goat Magazine, *April/May, 1997, and we wish to thank the author, Ed Jackson, and Roylyn Coufal, the publisher, for allowing us to use this material.*

"Fainting" Goats

You may have heard about them—the goats that when they are startled fall over in what looks like a faint. They really exist, and they really do fall over. However, they do not really faint. These goats carry the recessive gene for Myotonia Congenita that causes something like a reverse adrenaline action in the muscles or a stiffening when the animal is startled. The reaction can range from mild (causing stiff legs and a hobble) to extreme (where the goat, according to Ed Jackson, "locks up" and falls over). The reaction can last five to ten seconds after which the goat is fine.

Goats that "faint" have been called "Fainting," "Wooden Leg," "Stiff Leg," and "Tennessee" goats. Their history seems more like legend. In 1880 a man named Tinsley came to Marshall County, Tennessee, with four goats exhibiting the fainting trait. He worked on a local farm and when he left, the four goats stayed behind. These four goats were the first of the goats that are now recognized by the above names.

According to Lisa Johnson, her Tennessee goats are aseasonal breeders with a common 200% kidding ratio. They are easy kidders and excellent mothers. They also show some genetic resistance to stomach worms. They are hardy animals, converting extra feed into muscle and mass and make do on little supplemental feeding to maintain their condition.

The Myotonia trait results in the building of heavy muscling which makes them an excellent meat animal. Offspring of crosses may show the same "meaty" traits of the Tennessee parent. And the first and second generation of crosses to other breeds usually do not carry the "fainting" trait.

In 1988, Tennessee goats were put on the American Minor Breed Conservancy Watch list, as there were fewer than 10,000 of these goats worldwide, most in the United States. This fact has caused breeders to try to conserve the Myotonia trait (what makes this breed so distinctive), and in the past ten years, Tennessee goats have been crossed on dairy, Pygmy, Nigerian Dwarf, and brush goats in order to conserve the "fainting" trait. Therefore, you will see goats with this trait that resemble the breeds they were crossed on. Recent efforts have favored extreme stiffness and small size for the exotic pet market.

Those breeding these goats for the meat market find them an excellent meat animal with deep, wide muscular bodies. When crossed on the Boer goat, the kids are larger than Boer kids and more uniform in size and build.

According to their breeders, Tennessee or "fainting" goats have great potential both for those who want to raise and conserve purebreds and for those who want an excellent marketable meat animal.

International Fainting Goat Association

The IFGA registers fainting goats in one of three categories: Premium, Heritage, and Regular. These categories are based on the goat's degree of Myotonic State (full down or wooden leg) and lineage.

Their website

http://www.faintinggoat.com

has information about fainting goats plus show results, a member directory, registration rules and applications, a newsletter, breed standards, a list of defects, pictures, and more.

Cashmere Goats

We wish to thank Linda Fox, editor of CashMirror, *the journal for Cashmere producers, for providing the following information on cashmere-producing goats.*

Cashmere Goats

The Cashmere goat is not a breed of goat, but a type of goat. There is no such thing as a purebred Cashmere goat. A more correct term would be "cashmere-producing goat." Cashmere is the soft, downy undercoat grown by the goat during the winter and shed out in the spring. Most goats, including dairy breeds, produce some cashmere. A goat raised for cashmere production is one which produces a marketable amount of cashmere.

History of the Cashmere-Producing Goat

Cashmere goats have been raised for centuries in China and other countries. Today, most of the world's supply of cashmere comes from China, Outer Mongolia, Afghanistan, Iran and India. In the 1970's, it was discovered that feral Australian goats produced cashmere as well. Cashmere goats in the United States are mostly descendants of these feral cashmere-producing goats imported in the 1980's from Australia and New Zealand. Some of these have been crossed with Spanish (Texas) meat goats and other breeds.

Characteristics of the Cashmere-Producing Goat

Cashmere goats vary in size, color, coat length and ear shape, and most breeders do not remove the horns of their goats. Cashmere goats do well in a variety of climates. Minimal shelter is required in cold or wet climates. They are hardy and disease-resistant, requiring little care. The cashmere is harvested by shearing, in late January and early February or combed from the goat (as the Chinese do), when it starts to shed in early spring.

Cashmere goats are seasonal breeders who are bred from July to December to kid in the spring. Multiple births are common and kidding does not require human assistance.

For more information, contact a Cashmere producers' organization or *CashMirror*, the publication of Cashmere producers.

The following information was provided by Janet Hanus, Monmouth, Oregon, and we wish to thank her and the Pygora Breeders' Association for allowing us to include it here. For more information about Pygora goats, please contact the Pygora Breeders' Association directly or visit their web site at, http://www.pygoragoats.org.

The Pygora Goat

by Janet Hanus

The Pygora, primarily a fiber goat, is a very versatile goat in that it can produce not only fiber, but also milk, pelts and meat. It produces three recognized fleece types: fiber similar to very fine mohair (A-type), cashmere (C-type), and an intermediate fiber with characteristics of both (B-type). These fleeces are sought by handspinners, who like this very fine, unique fiber.

History of the Breed

The Pygora is a hybrid goat, resulting from crossing an Angora goat with a Pygmy goat. Katharine Jorgensen, a handspinner, developed the breed in the early 1980s, in an attempt to obtain a silver grey mohair fleece. Her theory was that the Pygmy would contribute not only color to the normal white Angora fleece but also the cashmere-like down characteristics found in the Pygmy undercoat. She did not get that fleece in her primary F1 cross, but as she continued to breed, she developed unique fleeces that she really liked to spin. Others joined her in breeding these goats, and in 1987 the Pygora Breeders' Association (PBA) was formed. Breed standards were defined and a registry developed. Pygora goats can now be found all over the United States.

Breed Characteristics

The Pygora is a double-coated, medium sized, well-muscled goat, as long as it is tall. The female's average size is about 22 inches at the withers; the male averages 27 inches. Adult female goats must be over 18 inches in height; males 23 inches in height. All Pygmy colors and their dilutions, as well as black and dark red, are acceptable.

Pygora goats are alert, curious, friendly, cooperative and easy to handle. They are easy keepers and produce kids, milk and fiber on a diet of hay, grain, browse, clean water and minerals. The Pygora seems to be able to thrive in many different climates. Their housing needs are simple; a dry place to eat and sleep, with adequate pasture for each goat, is all that is necessary.

Fleece Characteristics

Not only does the Pygora goat produce three distinctive types of fleeces, but these fleeces tend to hold their fineness with age. This is a characteristic that makes this fleece valuable in the handspinner's market. The three types of Pygora fleeces defined for these goats are characterized as follows:

Type "A"- a long fiber averaging 6 + inches in length. It drapes in long lustrous ringlets. It may be a single coat, but a silky guard hair is usually present. The fiber is very fine mohair-like, usually less than 28 microns. The handle should be silky, smooth and cool to the touch.

Type "B"- a blend of fibers with characteristics of both mohair type and cashmere type fleece. It's usually curly and should average 3 to 6 inches in length. There is an obvious guard hair. A second silky guard hair is also usually present. There should be luster and the handle should be soft and airy. The fiber should test, on average, below 24 microns. The fleece color is usually lighter than the guard hair color.

Type "C"- a very fine fiber, usually below 18.5 microns and can be acceptable as commercial cashmere. It must be at least 1 inch long and is usually between 1 to 3 inches. It has a matte finish and a warm, creamy handle. It must show crimp. There is good separation between a coarse guard hair and fleece. The fleece color is usually lighter than the guard hair color.

Registering Pygoras

In an attempt to control the genetics of this breed, the Pygora Breeders' Association registry still accepts original crosses between the Angora and Pygmy, (F1), as long as the parents are registered in their respective organizations: either the American Angora Goat Breeders Association (AAGBA) or the National Pygmy Goat Association (NPGA).

To register offspring from two Pygora parents, both sire and dam must be registered with PBA. Three-quarter crosses are allowed (i.e. breeding a Pygora back to an Angora or Pygmy), as long as the offspring have an acceptable fleece. The Pygora breeds "true."

Pygora Fiber (from the Pygora Breeders Association web site)

Pygora fiber may be spun and then knitted, woven or crocheted. Because of the fineness of the fiber, it spins into a lovely yarn that is soft enough to be worn next to the skin. Items such as baby garments or luxurious shawls are well suited to Pygora yarn. Pygora also felts beautifully and locks of Pygora may be used to create wigs, beards or novelty toys. Pygora pelts make wonderfully posh rugs or chair accessories. Thus, Pygora fiber is fast becoming crafts persons' and fiber artists' preferred choice for any number of diverse projects.

Hints

"Do you know how many years most people own goats? Three years! Seems like an awful short amount of time. Those who have goats longer seem to involve the entire family, invest in work-reducing equipment and routines, keep goat numbers under control, and don't try to make their hobby replace their regular job."

"The best way we've found to get children interested in raising goats is to take them to a nearby fair and let them talk with other children who have goats. If they catch 'goat fever,' the next step is to buy them a few good books, ones that are easy to read and have lots of pictures. For young children, try some of the story books that talk about goats or a goat coloring book. Be careful not to choose a book that shows goats eating tin cans, etc. Find out about local 4-H groups or goat clubs where your child can visit, and be frank about the amount of time and work needed to take good care of an animal."

"I keep my goat coffee mug at work. People always ask me where I got it, whether I have goats, where they can get one, etc. I give goat stuff for gifts and send Christmas greetings on the goat note cards. Using goat products is a great way to advertise just how wonderful goats are."

"Raising goats is really hard work, and it's pretty easy to get burned out. It's important to make time every day to enjoy your goats. One way to do this is to divide up chores so that everyone has some chore he or she really enjoys doing. I really enjoy milking, so it usually doesn't seem like work. Others like to feed kids. Then spend time just sitting and watching your goats, or go in the pens and play with them. You don't have to spend every minute caring for them."

"Goats get bored too! Give your goats toys to play with. You can make simple teeter-totters from surplus lumber and jumping platforms from spools. Then you can enjoy watching them jump, dance, spin, and play."

This article originally appeared in the Dairy Goat Journal *in 1996. We wish to thank Dave Thompson, the publisher, for allowing us to reprint it here.*

To Each His (or Her) Own

by Joan Vandergriff

I often think back to how it all began, how I got into goats at all, and how I decided on the breeds I raised—first Toggenburgs and last LaManchas. Thinking back on how these choices were made, parts of my story are probably not so very different from yours.

My first goats were Molly and Charles. Molly, an Alpine-Nubian cross, had shades of brown and lavender across her flanks. Charles was a huge, full-horned Alpine-like specimen. I found these two goats in a newspaper ad, which, of course, didn't mention Molly's habit of leaping over fences, her udder swung up over her back.

Our goats were going to live on land we owned near our house. Land that was unfenced. Land that had no water on it. Land resplendent with poison oak, brush and briar. Armed with sturdy ropes, we dragged our goats onto our land, tied them to trees, supplied them with water and expected them to prosper.

Obviously, we knew nothing about keeping goats. The first milking made that fact obvious even to us. We had attempted to milk a goat once before, at a fair, where a breeder let us squeeze out a dribble. Ready to milk Molly, I confidently sat on the ground, put a bucket underneath her and reached for her two enormous teats. Feet and bucket flew, and after many tries, tempers short, my husband wrestled Molly into a full-body hold while I tried to milk. No luck. We eventually gave up, muttering to ourselves something about our "gallon-a-day milker, ha!" Acknowledging our sucker status, we concluded that while our Molly was indeed a female, she was no milker, she must be dry. The next morning, her distended udder convinced us that she probably wasn't dry, and she eventually decided that putting up with my milking was better than her overfull udder. We had milk. We had goats.

And so it went, for a few months, until breeding season. Charles got more smelly, and we got increasingly tired of his "bucky" antics. We gave him to someone who needed a buck to put in with their does, and we went looking for a buck for ours. Oh yes, by now we had increased our herd to a few others of nondenominational breeding. We located a local breeder and made arrangements to bring Molly and the others to be bred. The moment arrived. The breeder pulled out a young Toggenburg buck for our Molly (an obvious Alpine-type doe). "Your doe looks most like a Toggenburg," the breeder said, "so why not breed her to a Togg?" Sounded okay to us, though later it was clear that he had only Togg bucks.

And so began the herd of Toggenburgs later known as Dionysius. Why Toggs? Ignorance. Luck. My experience in choosing a breed is not, I think, unique. Most of us make this choice before we know enough to make an informed decision.

Only later did other factors reinforce the original choice of breed. In our case, there were not a lot of Toggs in our area, and we rather liked that. It gave our goats built-in attention. If someone local wanted to buy a Togg kid or breed a Togg doe, they came to us. At local shows, we had to worry about making the show official, not necessarily who would go grand champion. And let's be honest. While it's great to win against formidable competition, it's also great to be a big fish in a small pond. We got a lot more exposure by not having great numbers of other breeders to compete with, and this made our rise in the Togg breeding world more rapid.

Other reasons for choosing Toggs then included consistency. Toggs seemed a bit more consistent to breed than other breeds we witnessed our friends raising. The old nursery rhyme, "When she was good she was very, very good, and when she was bad she was horrid," seemed most appropriate for both Togg bucks and does. It made culling pretty easy.

Our Toggs were also great milkers, with a Breed Leader and other Top Tens among them. Since, like most goat owners, we wanted our goats to pay for themselves, we sold our milk, and the more they milked, the more we made on our milk.

Our Toggs were easy-keepers compared to other breeds, kidding easily, requiring less than heroic measures to keep them alive and well, reproducing themselves in great numbers (we had quintuplets once!). In addition, they were very easygoing and adaptable. All of these were factors that kept us with this breed.

Much of the frustration inherent in raising goats comes from a mismatch between a breed's characteristics and the owner's personality. When it's time to decide which breed to raise, it's important to first choose a breed with traits that are compatible to you, and then breed and cull within that breed for these traits.

Many goat-breeder friends bought one breed then another, trying to find just the right match. But as I watched, it always seemed that they stuck with a breed because of one animal's traits, not the breed's. Old "Pixie" was an excellent milker, mother and champion. However, every other offspring and breed member had milked very little, shunned their kids, and invariably placed near the bottom of the line. No wonder these people were unhappy with their goats and bounced from breed to breed.

The same characteristics come into play within a breed. We bred and bought Toggs that quickly found new homes, mainly because they violated one or more of the characteristics important to our goat husbandry philosophy. Poor milkers, bad actors, and frail individuals were shipped down the road quickly, no matter how much we loved them or their bloodlines.

Every once in a while, we'd venture into another breed— a lovely Nubian doe, a few LaManchas. They too, had to live up to the same goals we had set for the Togg herd; consistency in breeding, showing, health, milking and behavior. They were shown the door for even less serious violations than their Togg herdmates', simply because it was so difficult to develop and maintain two distinct breeding programs.

Most of us don't seriously set out to choose a breed in a systematic way, and most of us don't learn much from our first experiences. In my case, I didn't use anything I had learned from raising Toggs to help me choose a breed the second time around—the same mixture of ignorance and luck played their role.

In choosing a breed, first impressions are important. I liked both the Toggs and LaManchas right off, and I liked the specific individuals of these breeds too.

But I considered other factors also. The most obvious, though perhaps not most important, was the popularity of the breed. In both cases, the breed numbers were low compared to other breeds. Since we were interested in marketing our animals nationally, it was helpful that there was little local competition and only a few outstanding national competitors. Marketing our animals, once we had something to offer, was much easier than if we had raised a more popular breed. We got "more bang for our buck and hard work" with these breeds.

Factored into our decisions were the general breed traits: the

Toggs were hefty milkers and our LaManchas were not, which suited our needs at the time. Both breeds were sturdy, self-reliant types, which suited our lifestyles. And the overall consistency in breeding and behavior allowed us a more relaxed attitude and more rapid advancement in the breed.

What parts of my story mirror yours? If you're like the majority of breeders, the part about ignorance and luck. You can't do much now about your original choice of breed, but if your goats are not living up to your expectations, you might want to do a reality check. We're inundated with information about how to cull for conformation traits, but we rarely think of the other, usually more important aspects of breeding goats, those that make breeding goats a joy or a nightmare.

Think honestly, about what's important to you as a breeder. Don't hesitate to acknowledge the less than sterling but truly important characteristics that might apply (in my case the big fish/small pond syndrome, the need for little milk). Rank your needs in order of importance. Is lots of milk most important to you or compliant personalities? If you want lots of milk, then you may have to put up with some poor behavior, or if good behavior wins out, jettison the two-gallon milker whom you just can't stand. You'll never know what to cull, though, if you don't first know what traits are most important to you.

Consider whether the problems in your herd are problems with individuals or ones inherent in the breed. Some breeds do have traits that certain breeders find objectionable. You may just find that you and your breed are not compatible. But be careful here not to indict a whole breed for the sins of a few individuals.

Goat families also have traits that pass from generation to generation, so if you find a few animals with an objectionable trait, look to their backgrounds. Is there a common dam, sire, grandparent? If so, consider culling those animals that share that commonality. Most important, if you find that a buck is passing on undesirable personality traits, use him the same way you would use a buck that passes on an undesirable conformation or production trait—rarely or not at all.

Conversely, recognize when a sire is passing on desirable personality traits, and use him on lines that need an "attitude adjustment." A good, adaptable personality is a marketable trait.

As part of your breeding program, then, know what personality traits are important to you. Acknowledge those traits that you just can't live with, those that drive you batty on occasion but are not worth culling for, and those that make your herd a pleasure. Armed with this information, consider each animal in your herd and make some hard decisions. Breed your animals using the same criteria. Maximize the positive personality traits and cull animals with obnoxious ones.

Finally, when you sell an animal, be honest with the new owner. Acknowledge that Petunia, though an excellent milker, is impossible to catch at milking time, or that Manly must live alone since he attacks both goats and people. Consider disposing of an animal with a truly horrendous attitude rather than selling it to another breeder. Unlike many conformation faults, poor personality traits may be hard to see right off, and we owe it to other breeders and to the breeds we are raising to intensify positive qualities and reduce negative ones.

We follow this philosophy when breeding for conformation and production traits. Now it's time to use these same criteria when it comes to breeding for personality characteristics and those traits that can make having goats such a pleasure or nightmare.

"Good News, Bad News"

This article originally appeared in the Dairy Goat Journal *in 1996. We wish to thank Dave Thompson, the publisher, for allowing us to reprint it here.*

Good News, Bad News
by Joan Vandergriff

We've all heard the good news, bad news joke: The good news is you're going to heaven; the bad news is you're going today!

Goat owning is a lot like that. The good news: Today, goat numbers are going up. The bad news: History tells us that sometime in the near future goat numbers should go down.

Interestingly, the increase in goat numbers is probably influenced by increased demands for goat products. However, a decrease in goat numbers may not be so closely related.

What accounts for this cyclical trend in goat owning? There certainly are sociological and economic influences which come into play.

Remember the "back-to-the-land" movement of the 1960s and early '70s? Then, thousands of people flocked to acreage to try to make their lives self-sufficient. Crammed onto as little as a half acre, they'd have a huge garden and orchard (taking care of family vegetable and fruit needs from apricots to zucchini), a compost pile and trash dump (no need to have pricey trash pick-up), lambs and pigs (sometimes a few steer instead—good material for trading to other back-to-the-landers who raised sheep and pigs), and of course goats. After all, every self-respecting homestead (because that's what we called 'em then, not farms) had to have a fresh milk supply, and a milk cow was too, well, you know, only serious farmers milked cows.

So that was the picture in the '60s and '70s. Lots of people working very hard to raise their own food, make soap, weave cloth, and milk lots of goats. Few people could keep up this lifestyle very long, especially with the lures of the big city calling from newspapers, radios, and televisions.

When people stopped to figure what a pound of homemade butter (substitute cheese, apricot jam, zucchini bread, or other product) cost in terms of time, back-breaking work, the impact on their family, and saw high school or college chums zooming around in new cars, going away on vacation, buying goat cheese in the gourmet markets, it soon became hard to justify the lifestyle. It was hard to keep back-to-the-landers down on the homestead, and pretty quickly goat numbers declined.

If we look at the economic side of things we see that when the economy is bad, people not only cut back on spending, they try to provide for themselves much in the way back-to-the-landers did. (Smartly, they usually do this on a much smaller scale.) So in tough times, we see people with lambs, pigs, steers, and the mandatory goats in their backyard. These are people who in better times wouldn't normally think of keeping livestock for their own family needs. And then what happens? Tough times eventually lead to better times, and when families find that their hard work with their goats no longer helps them financially, off to the sale barn the goats (and other critters) go. Again, a decline in goat numbers.

But if we disregard the sociological and economic influences, we are left with the very nature of goat-owning. It virtually dictates a cyclical trend. The increase is easy to understand: Goats sell themselves, so there will always be people who, by virtue of the goats themselves, want to raise them. As to the decline, let's look at a typical example.

Most beginning goat owners don't have any idea of what raising goats entails. They see goats at the fair or at a farm, they look cute, and soon they've bought a few. They've never tasted goat milk, goat cheese, or goat meat. Nor have most new goat owners raised livestock previously.

But soon, like the proverbial rabbit, a few goats become ten goats become twenty-five, and soon we're talking a huge herd—all in the span of two to three years. Trust me, the math works!

Having lots of goats means lots of work. Not only the daily milkings, but barn cleaning, maintenance, making goat products, correspondence with potential buyers (if you plan on selling goats), perhaps showing and other chores too numerous to name. I used to have a recurring nightmare of coming into the barn and seeing not goats but huge gaping mouths, demanding feed. You've probably had similar dreams—huge udders waiting to be milked, screaming babies waiting to the fed, fill in your own personal demon.

Lots of goats also mean lots of feed and other costs (equipment, veterinary bills, maintenance, etc.). The costs soon increase way beyond what we'd normally spend for family entertainment or hobbies.

"Good News, Bad News"

Goat owning may start out as a family endeavor, but more times than not, responsibility eventually falls on one person. Eventually, such responsibility usually brings with it ill feelings. One family member is out in the barn at all times of the day or night while junior or spouse isn't. A family hobby begins to divide the family, leading to family disharmony if not divorce.

Within a short time (remember, the average goat owner stays in goats for three years), keeping goats doesn't seem like such a good idea. It's back-breaking work, it's financially unrewarding, and it's hard on family relationships. Now it's decision time. Either get rid of the goats (which many people do), or decrease the work, increase the financial return, and/or improve the division of labor.

At this point, some people decide, illogically, to go into a goat-related business. They may choose selling milk, making cheese or other dairy products, selling goat meat, or some other money-making endeavor.

Soon, not only are they shelling out tons of money, putting in long hours, and arguing with their family about the goats, but they're also shelling out tons of money, putting in long hours, and arguing with their family about their goat-related business. Again, such situations lead to family disharmony if not divorce. End of business, and off go the goats to the sale barn.

Call me cynical, but over the twenty-five years I've had goats, I've seen this scenario played out repeatedly. But does this cyclical trend have to be? Well, we can't do much about the sociological or economic influences, but we can do something about the nature of owning goats.

To stay in goats over the long haul, we must find ways to decrease the work and financial outlay and keep the family as a whole committed to this hobby.

The first way to do this is to keep numbers to a minimum. Defy the "breed-like-a-rabbit" syndrome.

Keep only as many goats as you can handle (the number will differ depending upon if you're working by yourself, with family members, or with other help) and still have a life besides goats. This will help in many ways. You'll work less, you'll spend less, you'll argue with your family less, and you'll enjoy your goats more.

Don't think that by increasing your herd size you'll make everything inherently better. You'll probably gain some efficiency and per head cost reductions, but for most people, more goats mean more work, more costs, and more disharmony.

Keep only as many goats as you can afford. Since owning goats is a hobby, consider having goats the same way you consider your entertainment budget. Just the way you decide that you can rent only four movies a month, decide how many goats you can afford to keep.

Don't try to turn your hobby into a paying job in order to justify a large herd. First, most people have never run a business nor have they worked for themselves. Whatever the business, being successful will depend upon your knowledge of business planning, marketing, accounting, and a number of other skills that most of us do not naturally have. And statistically, the results are preordained since about 90% of all new businesses fail.

Next, demand efficiency in all your goat chores. Be process-oriented. Analyze your milking, feeding, cleaning, and other labor-intensive chores to minimize effort and maximize efficiency. You must do this consciously. Put pencil to paper. List your chores and decide on the most labor-intensive parts. Then, chore by chore, decide how to decrease your effort. Decide whether you need better equipment (and whether you can afford it), better processes, and/or more help. Be specific. Then implement these changes. If you've made your chores as efficient as possible and you're still suffering overload, you'll need to reduce numbers or get more help.

To improve family relationships, you must begin with an honest approach to family involvement. This means real discussions with everyone about **who** really wants to be involved with the goats, and **how much** of a commitment each one is willing to make day in and day out. This can't be the same type of commitment made when you bought the puppy—"Sure I'll feed it and walk it," says Jimmy, and then Dad or Mom is left with the job.

Everyone must be clear about what the commitment really means. To do this, make a list of everything expected of each family member. Putting it in writing helps by making the commitment real. Some people make a family contract and have each family member sign it.

At set intervals and as needs change, revisit your "family contract" and "renegotiate" parts of it that need changing. Hold a family meeting immediately if one member isn't participating.

Discuss where the problem lies: is it with specific parts of the commitment, e.g., has cheerleading, football or changed work hours interfered with afternoon kid feeding, or is it the overall commitment itself, e.g., your spouse will go mad if he or she ever milks another goat!

Consider the total picture (which means more than just goats). For example, if every family member except you wants out of their commitment, you'll either need to get rid of the goats or scale down so that you can retain your more important role as family member.

When we look at what owning goats really entails, it's pretty obvious why most people stay in goats for only three years, and why those that stay longer often work harder, spend more money, and suffer from more strained family relationships than they should.

However, the cyclical nature of owning goats is not a necessary condition. We can continue to enjoy our goats and have a real life too—if we keep goat numbers down, introduce efficiencies, allow the family members to be involved as they want to be, and remember that for most of us owning goats is a hobby. Like other hobbies, owning goats should fit within the framework of the family, giving pleasure and teaching responsibilities, for as long as family members continue to reap these rewards. In this way, we can keep goats as part of our lives for a lot longer than the average three-year cycle.

General Management: An Overview

When we raise any kind of animal, while we do not necessarily think about it, we want to make sure that we do everything we can to ensure its health and well-being. We all know that if we start with strong, healthy stock, we are more likely than not to raise good offspring. However, just having good stock to begin with is not enough. To succeed in any kind of animal husbandry project, we must practice good management of our animals. In fact, in many cases, good management in the end may be more meaningful than starting with the very best stock. (We have all known or heard of people who spent thousands of dollars on stock, only to have them fail to thrive because of lack of knowledge and poor management.)

Because of the nature of goats themselves (they are so cute and seemingly easy to keep), most first-time goat owners tend to take on their goat project without much background in goat husbandry specifically or animal management in general. To begin with, we are so enthusiastic to get started that we often do not do our homework before taking on the responsibilities of raising our animals. We learn by "on-the-job-training," mostly from talking with others and through our mistakes. Eventually, most of us can use our good judgment and develop a "feel" for how to overcome problems.

However, as new goat owners, there are a few general principles that we should consider as we begin our goat projects.

First, **good information is the key to good management**. We need to know what to feed our animals and why, how to shelter them and keep them safe both inside our barns and out in our pastures. We need to know how to evaluate the stock we have, where to go for improvement, and what we have gained (or lost—yes that happens!) through our breeding efforts. Knowing how to keep our goats healthy and what to do when they are not are crucial aspects of goat keeping, and knowing when we are in over our heads is vitally important.

How do we gain this knowledge? Most novices have "experts" they can consult in local goat clubs or just neighbors, but you need to be able to weed out the "old wives' tales" from accurate information. In many parts of the country where there are active goat clubs or 4-H projects, you'll find informative workshops and dairy goat days where you can attend seminars and talks led by people with years of practical experience and experts with specialties in areas such as health, genetics, breeding techniques, etc.

Your local county extension agent may be another source of information. He or she can tell you about other goat owners in your area, activities that might be of interest, production and evaluation programs you can participate in, and local particulars (e.g., common toxic plants or weeds, average protein levels in local hay or forage, diseases that are endemic, etc.).

If you have children with goats, you will want to investigate whether there is a local 4-H club that has a goat project. In some cases, you will find that in order to have your children participate in a 4-H goat project, you will have to initiate it. While such projects tend to begin with the "blind leading the blind," both you and your children will benefit enormously from this experience and learn more than you ever thought you would know about goats!

Even if you do not have registered goats, you should consider joining an association and contacting the club(s) that represent the animals you have (e.g., the American Dairy Goat Association, a Boer goat association, the National Pygmy Goat Club, the American Harness Goat Society, etc.). These associations are the source of excellent information specific to your animals, including programs and expositions that you can participate in, and their mailing list will give you names of local people with like interests.

Investing in a few good books is always wise. Your "goat library" should include a good book on general goat husbandry (something specific and comprehensive but not so in-depth that you need a Ph.D. to read it), a goat health book (again, there are one or two inexpensive but excellent books that fill this need), and if you are interested in pack or harness goats or other specialties such as making cheese or soap, then you will want to own a book that focuses on your area of interest.

If you own a computer, then you have the whole world (web, that is) to browse through. Begin by using your favorite search engine to search on "goat," and you will be amazed by the amount of information you will find and the links you will be able to follow. Download files for later reading and print out those that are of special interest, and you can begin putting together your own reference library that suits your specific needs.

General Management: An Overview

There is nothing more frustrating than trying to remember what feed mixture increased your milk production or what month your best doe normally comes into strongest heat or when you vaccinated the middle batch of kids. Much of the information you need to practice good management comes from recording the events that take place day to day in your herd. This means that you need to keep track of information important to the health, breeding, and general management of your herd.

From our experience, the best gardeners, animal breeders, and ceramic artists are those who keep extensive records. As goat owners who are in a barn usually twice a day, the easiest way to do this is to keep a calendar in the barn and write down **every** important piece of information you need to remember, **every day**. A calendar with large blocks to write in works well. At the end of the year, you can transfer important information on to the new year's calendar. Calculate kidding dates, and write the resulting dates on the calendar. Keep track of vaccinations and antibiotics administered and when milk will be good for human consumption again. Track milk production day to day. Write down any health problems and what you did to "cure" them. Make it a habit to check your calendar each morning before you begin chores. Then you will know when Molly is likely to need breeding, Polly her tetanus booster, and Rolly his mineral supplement.

If you have a larger herd and need more room than the daily blocks of a calendar can provide, consider keeping a "diary" in your barn along with your calendar. Use the calendar to note important dates, but write more extensive notes in your diary. Be sure to date each entry. Our herd diary includes pages of breedings (who, to whom, and when), dry dates, kiddings (who, how many, a description of each kid, any abnormalities, kidding problems, or other notes), tattoo information, show wins, sale information (deposits taken, balances paid, kids shipped) and any other information deemed important enough to be recorded. Keeping this comprehensive diary means you will have historical information at your fingertips, and you can use this information for years to come.

It really is true what is said about **cleanliness**. When you raise animals, you can never have a barn, milking parlor, pasture, kid, doe, or buck pens that are too clean. We have all been in barns where the odor from built-up bedding makes our eyes and nose run.

Remember that we only have to be in our barns for a short time each day. Our goats, however, live there all day, every day. If your eyes or nose is irritated, think about what it is like for them.

Nothing is less healthful than bringing goats up in a damp, smelly environment, and you will find that your animals will grow faster, have more kids, milk better, and just be in better overall health if the barns they live in are kept clean. This means that when bedding starts to build up, you clean out the barn and put out fresh bedding; if areas tend to get swampy, you try to move the stream of water away from the barn by trenching; if your barn is airtight, you put in some cross-ventilation or fans or anything that will allow air to circulate and fresh air to be exchanged for "barn air." The first whiff of ammonia should send you scurrying for rakes and fresh bedding.

It is hard to have clean milk in a dirty milking area. Think about dairies you have seen pictures of. Everything is white for a reason. It connotes cleanliness (and reminds us that cleanliness is essential), and you can see any dirt in a white environment. While you do not need to paint everything white to be clean, strive to make your milking area as clean as it can possibly be, and do the same for anything that milk touches (milking equipment, milk storage bottles, kid feeding equipment, etc.)

Clean out feeders and waterers regularly. Remember that goats like to climb in and on feeders and waterers and are especially susceptible to worms that such climbing can introduce. Clean water is just as important as clean, good quality feed. Make sure that goats have ample water and that their water source remains clean. Scrub water buckets as needed so they do not get algae build up. Check waterers often in hot weather to make sure there is an adequate supply of clean., algae-free water.

Learn to recognize hazards to your goats and you in your barn, milking area, and pastures. Dirty conditions are hazards, but while your barn may be clean, make sure your pastures are free from manure build up and other unclean conditions. Nails and ripped fencing can cause serious puncture wounds to animals and humans alike. Make sure that you remove any woodpiles before you let your animals out into a pasture or a barn area. Do the same with any debris, no matter how benign it seems. An old car can trap a goat underneath, its fender can break a leg caught behind it. If goats are going through fencing, they'll eventually injure themselves. Button up those holes to keep your goats in and keep them safe.

There are probably lots of other principles that apply to raising goats. But these are some of the basics. The longer you have goats, the longer your list will become, and your list, while sharing commonalities with other goat owners', will reflect those parts of goat-keeping most important to you, your herd, and your own environment.

Basic Feeding Plans

There are as many different "right" ways to feed goats as there are breeders. Consider these contrasts:

A New Jersey breeder who had very good production records for her dairy goats followed a "challenge" feeding program. Potentially high-producing does got as much as four to six pounds of grain a day before they freshened. Milking does got a pound of grain for every two pounds of milk they produced.

A California breeder with equally good production records fed alfalfa hay free-choice but limited grain. Dry does got no grain at all. Milkers got three pounds of grain a day, no more.

Another breeder fed an all-in-one ration that was about half cottonseed hulls, free choice, and no hay at all. It worked for them, but it might be inconvenient or expensive for others.

All of these goat raisers were convinced that their feeding plan was the "right" one, and it was — for their herd.

Here is a more usual feeding program:

Most animal nutritionists agree that a 14% to 16% protein grain mix works well for dairy goats. Make sure that the dairy mix you buy isn't dusty. Goats prefer no-dust grain that is glued together with plenty of molasses. Substitute a horse grain mix if your local dairy mix doesn't have enough molasses to suit your goats or if it is too dusty.

Hay is a critical factor. If you can get well-cured high-quality legume hay, it should be high in protein, and you will need less protein in your grain mix. If the best hay available is grass, that is just fine, but grass hay has less protein than legume hay, so you will need a grain mix with more protein in it. Make sure that your hay is not overly dusty or moldy. As with the grain mix, goats do not like dusty hay, and moldy hay can make them very sick.

In certain parts of the country, finding good hay is very difficult. Talk with other local breeders, and check with horse breeders and your local cow dairy, if necessary. Also remember that grass hay is usually less expensive than a legume hay, and it might be more economical (and more healthful to your goats) to feed a good quality grass hay and up the protein with grain and mineral supplements (which you might be feeding anyway).

Growing kids may get a pound of grain a day and all the hay

they can consume. Bucks are more variable. Depending upon their size, health, exercise, and amount of use they are getting, you might feed them anywhere from one to four pounds of grain a day and usually grass hay.

In the last two months of pregnancy, you want to put your does on an "ascending plane of nutrition." This means that you want to gradually increase their grain ration. Some breeders feed up to two pounds of grain a day towards the end of pregnancy. Again, watch the condition of your pregnant does. A thin doe, carrying a huge load of kids, the mature doe that milks a bunch will both probably need more grain.

Many breeders feed one pound of grain per three pounds of milk produced. They give less to fat does, more to thin ones.

Most breeders feed hay free choice, allowing goats to eat all they want. An adult goat will eat four pounds of hay or more.

Make sure that the hay feeders are clean. Goats will often put their feet in the feeders, increasing the possibility of contamination from worms. Goats also have the reputation as picky eaters. They will eat the leaves off the hay and leave the stems. Check your feeders to make sure that the hay that remains is leafy and palatable, not just stem. You do not need to throw away the stemmy leftovers. They make great feed for calves you may be raising on goat milk.

Those raising meat, hair or working goats (goats used for packing or pulling) may rely more on forage and you will need to try to balance your goats' ration with an appropriate grain ration. With working goats especially, you will need to keep a close eye on their condition and supplement their forage ration if you see they are looking a bit thin.

Water is as important a "feed" as grain or hay. Always provide plenty of fresh, clean water for your goats. They also need a salt block or loose salt. In addition, many breeders will feed a loose mineral mix either free choice or sprinkled on the grain ration. Such supplements may contain beneficial bacteria which are intended to improve digestion and increase feed utilization. There are other mineral mixes meant to balance the calcium and phosphorus in your goats' diet. Finally, you can feed your bucks a mineral supplement meant to help increase their fertility and potency and improve their overall condition.

Balancing Calcium and Phosphorus

You usually do not have to worry about a calcium or phosphorus "deficiency" in your feeding program. It is much more important to balance the amount of calcium and phosphorus in the total ration: grain, hay, and/or other forage. If there is too much of one mineral in relation to the other, your goat's metabolism may go awry, and illnesses such as "milk fever" may result.

To be absolutely sure that your ration is balanced, you will need to have your hay or forage analyzed. Your county extension advisor should know where this can be done, and he or she can show you how to calculate the calcium-to-phosphorus ratio of your goats' total feed intake.

Alfalfa hay is high in calcium and low in phosphorus. Grass hays and grain are high in phosphorus and lower in calcium. For dairy goats, the ideal mineral balance is a calcium-to-phosphorus ratio of 2:1 or 1:1. When feeding minerals, check the labels carefully. Some have a calcium-to-phosphorus ratio of 8:1 or even higher, which can really skew the ratio without your realizing it. You will then need to balance with other feed accordingly.

The Basics of Breeding Goats

Goats Are Seasonal Breeders

Depending upon the climate, goats tend to breed in the fall and early winter, and kid in the spring. Bucks may be ready to breed in mid-to-late summer, so you will need to be careful that they are housed securely if you do not want winter kids.

The Heat Cycle

Between August and January, your does should come into season every 18-21 days. They will stay in heat for varied lengths of time, some only a day, some three or four days. For those with longer heat cycles, there will be a time when they are more receptive to the buck than other times during their cycle. Many breeders allow their does to be bred repeatedly during the heat cycle to ensure conception. Doing this will depend on how much you need the services of the buck. If he has numerous does to service during this period, you will want to limit his service per doe.

Signs of Heat

Many does will indicate that they are about to come into season. They will be a bit more active, more vocal, more demanding of your attention. If you try to breed them, they probably will not stand or allow themselves to be bred. This pre-season activity may last half a day.

As the doe really comes into heat, you may see a mucous discharge from her vagina. As the heat continues, this discharge may turn more clear and stringy or runny. Usually when this happens, the doe will stand for the buck to service her, and midway in her heat cycle may be the best time to breed her naturally (using a buck).

As the doe comes out of heat, the discharge may change again, becoming more opaque and thicker, but this may be hard to tell as she may have a discharge from the breeding, also. Usually does will not stand to be serviced as they are going out of heat.

If you are using artificial insemination, you will probably be more successful breeding in the latter half of the heat cycle, as the egg is released late in the cycle and will stand a better chance of being fertilized if the doe is inseminated later in the cycle rather than earlier.

Breeding Goats

Breeding Young Does

Theoretically, young does are ready to breed by the age of four months. However, for most of your kids, you will want to wait until they are eight months of age. Many breeders use weight as a basis for breeding. Some say 45 pounds is okay, others 70. Remember that young does who are bred are not only growing kids but are growing themselves and need close attention to their feed ration and their condition overall.

Maturation of young does will vary by breed and genetic background, so you can not expect that a small-boned LaMancha may be ready to breed at the same age as a large-boned Saanen. Some people breed their young does to kid at one year of age; others hold all young does over as drys during their first year and have them kid for the first time as two year olds. Talk with breeders in your area about when they breed their doe kids, and you will probably get as many different answers as you have breeders. You will have to use your best judgment and base your decisions on the size and maturity of your own young does. Some studies have shown, however, that on average, young does held over to kid for the first time at two years of age never reach the production levels of those bred to kid as yearlings.

Check for Pregnancy

Do not assume that your doe is pregnant just because you have bred her a couple of times during her heat cycle. You should have written down on your barn calendar the breeding dates of all does bred. Be sure to check for the next heat cycle, and rebreed does that seem to have recycled. Some does will continue to cycle (and may even allow the buck to service them) even if they are pregnant. Over the years, you will get to know the individual differences of each of your does. In the beginning, you will need to keep a close eye on all your does.

Breeding Calendar for Goats

The average gestation period for dairy goats is 150 days or about five months. Use the calendar below to help you determine when your does will kid. (Obviously, you'll need to have kept good records of your breeding dates to use this calendar successfully.)

To calculate the kidding date, write in the breeding date and subtract the number indicated under the appropriate kidding day.

If a Doe Is Bred on (fill in date)		**She Should Kid On**	
Month	**Day**	**Month**	**Day**
July	______	December	______ - 3
August	______	January	______ - 3
September	______	February	______ - 3
October	______	March	______ - 1
November	______	April	______ - 1
December	______	May	______ - 1
January	______	June	______ - 1
February	______	July	______ 0
March	______	August	______ - 3
April	______	September	______ - 3
May	______	October	______ - 3
June	______	November	______ - 3

Example: If a doe is bred on October 12th, the breeding calendar indicates that she should kid on March 11th.

Artificial Insemination: You Are Just the Middleman

Goat owners have been using artificial insemination for breeding dairy goats for many years, but there are still breeders with large as well as small herds who think that artificial insemination (or A.I.) has little to offer. They could not be more mistaken. Artificial insemination offers outstanding opportunities for all goat breeders.

It is a wonderful aid in breeding programs. There are numerous sires in all breeds available through A.I., so you can choose one whose strengths will complement those areas of greatest need in your herd. You can use five different sires, if you wish, on five different does. You can pick just the genetic background you are looking for.

On the practical side, you can use bucks 3,000 miles away; you can even use bucks that are dead.

If you keep more than one breed and if you use A.I. exclusively, you will never worry about the wrong buck getting loose and breeding your best doe of another breed.

Other practical considerations? It is expensive and difficult to build good housing for bucks; they can break out, get unruly, and just be nuisances. Also, the initial cost of a good buck plus maintaining him are other expenses. Do remember that once you have used this buck on your does, you will have to find and pay for another buck to breed the daughters to. And of course, if you must travel to breed your does, you know they will come into season on the snowiest day of the year.

On the other side, it is costly when you begin to A.I. You need a nitrogen tank to store your semen in. But breeders have grouped together to share a tank or have rented space in another breeder's tank or have put their goat semen in the cow dairyman's tank down the road. The minimum equipment you will need to begin (once you have the semen and a place to store it) includes an inseminating gun and sheaths, a light source, two speculums, lubricating jelly, a straw cutter, and tweezers. For about $150, you can be in business. While this is a lot of money, remember all the expense you will be saving on the cost of a buck, housing, feed, travel, and so on.

The last worry you may have is actually getting the does settled using A.I. First, take a course given by a reputable A.I. technician. You can do this at an American Dairy Goat Association convention, plus A.I. technicians offer them around the country as they travel collecting bucks. Take the class, listen carefully, and try inseminating as many does as you can during the class; and, when you have the chance, A.I. your own and your neighbor's does. Most important, keep practicing as often as you can.

Like most other things, the secret to success at A.I. is practice: the more does you inseminate, the better you get at it.

Artificial insemination is an important way to improve your goats. Just look at the foremost breeders of dairy cows. Virtually all of them use A.I. as an important part of their breeding programs.

Remember, there is no big mystery to A.I. You're just the middleman.

Embryo Transfers

We are very grateful to Drs. Brian McOnie and Murray Flock of Embryogenics and Creekside Animal Clinic, Vernon, British Columbia for the following information on embryo transfer in goats. For more information and the very latest development in this technology, visit their website at http://www.creeksideanimalclinic.com.

Embryo Transfers

Embryo transfer (ET) is an advanced, but well established, animal breeding technology. ET encompasses the several procedures involved with the recovery of embryos from the uterus of a donor female animal, approximately 6-7 days after breeding. Good quality embryos may be transferred into the uterus of recipient ("surrogate mother") females at a similar stage of their reproductive cycle. Alternately, embryos may be processed and held frozen in liquid nitrogen for storage and movement for transfer into recipients at another time or place

Many benefits that arise through the use of artificial insemination (AI) also apply to the use of embryo transfer. AI allows the distribution of the genetics of elite males, each of which may each produce thousands of doses of semen. By contrast, ET favours the contribution that may be made by superior females. When ET is coupled with AI, genetic progress may be accelerated even more. Each ovary of a ewe or doe contains several thousand oocytes (potential ova or eggs), of which perhaps 30 may be released and fertilized over her life-time. Super-ovulation and embryo transfer allows us an opportunity to increase several fold the production of offspring from these females by making available for fertilization a greater number of ova which would otherwise not be used.

The size of sheep and goats, their seasonal reproductive behaviour, and aspects of their anatomy make the application of AI and ET more complicated than in the larger domestic species (i.e. cattle and horses). Those considerations have not deterred serious breeders in sheep and goat producing countries from adopting ET.

Why use embryo transfer ?

1. Genetic improvement - ET allows an increase in the proportion of the herd or flock derived from genetically superior females.

2. Bio-security - Specific biological aspects of embryos, combined

with special handling, processing and packaging procedures reduce the risk of domestic and international disease transmission.

3. The international movement of novel breeds - The costs of shipping, quarantine requirements and potential risk exposure are much lower for embryos than for live animals.

4. Salvage of genetic material - The production of clean or specific pathogen free herds or flocks from donor animals of high genetic merit, but low health status, is entirely possible. Use of lower quality female recipients to produce high quality offspring.

5. Storage of genetic material - Embryos can be stored almost indefinitely in liquid nitrogen and may be recovered from donors and transferred into recipients in or out of the regular breeding season.

Expectations and caveats

Approximately 25% of potential donors do not respond to superovulatory treatments. Pregnancy rates are generally, but not always, lower with frozen than with fresh transferred embryos.

Embryo collection and transfer in small ruminants involves general anesthesia and surgical procedures. Although modern anesthetics and surgical techniques reduce risk to donor and recipient animals, it should be appreciated by owners that serious and potentially fatal complications, while rare, may occur.

Pregnancy rates are variable and relate to donor, recipient, embryo and operator factors, many of which may interact, usually negatively, to influence the outcome. Clients thinking of an ET program should be prepared to follow directions explicitly. There are no guarantees and careful attention to detail is important to reduce the chances of a program being disappointing and increase the probability of it being remarkably rewarding.

Fly Control

The best control for flies is cleanliness. Cleaning pens and hauling out bedding and manure once a week will go a long way toward controlling flies.

Other methods that have proven effective in controlling flies without chemicals include hanging up lots of fly strips around the barn and milking area and setting out fly traps. Some breeders report that feeding livestock diatomaceous earth will cut down on flies breeding in the manure. Others have sent away for fly-eating larvae and report good results. Today, there is a great new product called "Fly Predator Insect Control." These tiny stingless wasps, when spread on your worst manure areas, lay eggs inside pest fly pupa, and the developing fly predators consume pest fly larva. The key here is to use these fly predators consistently each month until frost season. Fly predators will not kill existing flies, but will minimize development of new pest flies.

Using a good dairy insecticide will overcome the most stubborn fly problem if used in sufficient quantity along with a regular program of barn cleaning.

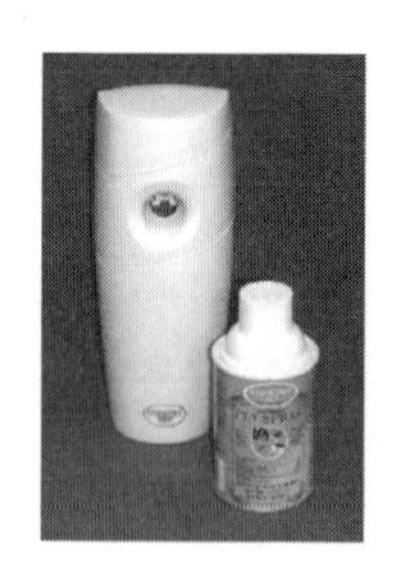

In our large barn, we set out automatic mist sprayers above each pen and in the milkroom. We monitored them to make sure they were working and raked pens weekly. We also hung fly strips where we found flies congregated at night. (The Sticky Roll Fly Tape Systems worked extremely well here.) This type of routine really got rid of our flies and kept them away. However, if we missed cleaning the barn or ran out of spray, the flies returned almost overnight.

So no matter which method you use to control flies, natural or chemical, you must also establish a routine cleaning schedule. And stick to it.

Overcoming Bad Habits

"I use teat tape on a doe that is a self-sucker. In the beginning I used lots of tape. Now, since she is pretty much out of the habit of sucking herself, I can use only a small amount of teat tape. The best part is that the teat tape doesn't irritate her skin the way other tape I tried did."

"We use a 'breaker bar' to stop goats from jumping fences. A breaker bar is a foot-long piece of half-inch steel rod that hangs at knee level from the goat's collar. When a goat wearing a breaker bar tries to jump it bumps its knees, which discourages it from jumping. In most cases, once a goat learns not to jump in this way, you can remove the breaker bar."

And a variation: "We had a yearling Togg that kept jumping in feeders, walked along the top of the keyhole feeders, and broke the lights. I was going to take her to the auction because she was so destructive. We decided to try the 'trapeze' for jumpers (breaker bar). It took six weeks of wearing this device, and she doesn't jump now. Then, a bossy, bratty doe started smashing other does in the herd. I put the trapeze on her, and, sure enough, when she would zoom off to nail someone to the wall, the bar would hit her knees, and she would stop dead in her tracks. A wonderful invention!"

Hints

"Goats won't drink dirty water willingly. To keep water clean, you can set your water buckets outside the pen on the ground or on a shelf. This works well with grown does. Small kids, however, being the devils they can be, may climb through the opening in the fence. If kids are mixed in with does, kids may not be tall enough to reach through the fence and get their heads in the water bucket."

"Goats appreciate warm, almost hot water in winter. They can drink a lot, and it won't chill them the way cold water may. This helps keep milk production up. By feeding hot water, you can water bucks and dry stock just once a day so that pails aren't left to freeze in pens. To keep water from freezing in buckets, try using a water bucket warmer; your goats will have water available 24 hours a day."

"The first thing we do every morning and evening at chore time is to 'count heads'; that way we know immediately if we have some problem. We also try to pick up on any goat who looks the slightest bit 'off'—standing off in the corner, hair on end, hunched up."

"Goats thrive on routine. Try to feed both kids and mature animals at the same time every day. Also, try to milk and schedule other activities (such as going out or coming back from the pasture) for the same time each day. Your goats will love you for it, and be healthier too."

"Our barns and pastures were such a mess, but we found a way to clean them up and keep them clean. We set aside one Saturday a month as clean-up day, and we planned nothing but clean-up for our Saturdays. In the beginning, clean-up took all day, but once clean, it took only a few hours to take care of problems that we couldn't get to daily. The trick is to set up a routine and stick to it. NO MATTER WHAT! A pizza for dinner also helps!

"Barns and pastures can really be dangerous. We make a habit of routinely checking for loose boards where inquisitive goats will stick their heads and choke, gopher holes they'll step in and break legs, protruding nails and fencing they'll snag themselves on, exposed wiring they'll nibble with electrifying results! It's probably impossible to 'goat-proof' a barn or pasture, but this type of vigilance helps cut down on disasters."

"Put some udder cream on the udder of a doe that's ready to kid. It helps relieve udder congestion, and it helps keep bedding and birth fluids from sticking to the udder."

"It's important to check the nitrogen level in your semen tank regularly. We try to do it the first of every month. We also check the level after we've moved the tank or before we remove a straw for breeding."

"To load goats in the back of a pickup truck, we put a bale of hay or straw below the tailgate. It makes a nice step up for them to hop on to before mounting up in the bed of the truck."

And to help those baby bucks: "Sometimes our baby bucks aren't tall enough to reach our grown does to breed them. We've solved the problem by backing our does up to a landing, dirt bank, or bale of hay or straw. The buckling can then stand up on his 'booster chair' and reach the doe. You need to make sure that when he dismounts he is careful not to fall off the step and hurt himself."

"At our farm, we think our goats should have Christmas too, so on Christmas Eve my husband goes to the local Lions or Kiwanis tree lot and buys up enough leftover trees to put one in every goat pen. He buys only natural trees, free of colorant or flocking, of course. This way we feel we're doing a little something for those more needy than we are, plus the goats just love the trees."

"For lightweight goat coats or to keep kids clean at shows, we use old t-shirts. They go over their heads and their front legs go through the arm holes. To keep them in place, working from the goat's back, gather up the bottom edge of the t-shirt and knot it over the top of the goat's back. We always keep a bunch of t-shirts in the barn cupboard or our show tack box."

"My Life (and Yours) As An Efficiency Expert"

This article originally appeared in the Dairy Goat Journal *in 1996, and we reprint it here with our most sincere thanks to Dave Thompson, publisher.*

My Life (and Yours) As An Efficiency Expert

by Joan Vandergriff

It seems that when one reaches midlife, there are two options: crisis or crisis management. You can wallow in your faded looks, extra heft, lessened staying power and . . . add your own qualities here . . . or you can acknowledge all of these, make any changes that need to be made, and move on. Midlife is a time for assessment and reassessment. A time to look at what you've done, what you'd like to do with the rest of your life, and—if you're like me and a little nostalgic and self-indulgent—what your life might have been if you had made other choices.

I regret very little in my personal life choices, but in my professional life I sometimes dream of other roads I might have chosen. I'd have loved to have been a superb athlete, a talented jazz singer or piano player, a daring race car driver. Closer to reality, one unrealized professional dream has been to be a meteorologist.

If I were to have based my career choice on what my personality is most suited for, my career choice should have been clear. I could have earned a Ph.D. as an efficiency expert without much trouble. My need for directness and clear thinking has caused its share of havoc with spouses and friends, but one area where my penchant for efficiency has served me well is in raising goats. I think at one time or another, we all need the advice of an efficiency expert in our goat-raising operations.

First, raising goats is labor-intensive to begin with. And since many of us are hobbyists, we don't always operate like a business, maximizing profit, minimizing expense. We usually don't count our efforts (e.g., our work) as an expense. Therefore, while we may not realize it, much of what we do and how we do it adds to our labor but does not bring much in return.

In my case, as our goat breeding operation developed, I began to see areas where we could make things easier for both ourselves and our goats. Soon I became the herd's efficiency expert, while my husband, bless him, was the person who figured out how to put my efficiencies into practice.

There are a few basic rules to follow when it comes to introducing efficiency into most operations.

#1: The old saw "A place for everything, and everything in its place," while it grated on us when Mother said it, is really the foundation of any efficient operation.

Leaving a piece of equipment to sit where you last used it is a fatal mistake. When you need it next, you will have to search in two places: where the item might be stored and where you might have used it last. And trust me, most of us aren't up to this memory challenge, and we often lose interest before completing our search.

To help keep track of your equipment, assign a place for each item. Group like items together. For example, keep all the hoof care items (trimmers, rasps, blood stop, etc.) on a single shelf. Tattooing stuff (pliers, digits, and ink) can be stored together in a plastic case. To help remind you where things are or should be, label the edge of the shelf or outside of the cabinet or case, using a label machine or simply masking tape.

Make sure your storage area is as close as possible to where you will be using the equipment. We kept our disbudding equipment in a cabinet above our baby pens. We made sure we had an electric outlet near there too. Since most of us try to find any excuse to put off the unpleasant chore of disbudding our kids, storing equipment in the house or in another barn or even a distant part of the barn feeds our willingness to procrastinate. Having items close at hand gives you that little nudge or reminder to do the job. If your tools are right at hand, when you have a few spare moments, you can fill them by disbudding a few kids: all you have to do is reach up into a cupboard, plug in the iron, and pull out a kid.

But just having a place to store your equipment isn't the whole answer. You must make sure that you return every tool or piece of equipment to its storage place. Do not consider an operation complete until you have inspected both your work area to make sure all tools and equipment are cleared away and the storage area to make sure that tools have been returned to their proper place. An added benefit: doing this also helps save money. How many hoof trimmers, for example, have you had to replace because you left them sitting where you used them last and they disappeared in the hay, bedding, or . . . ?

#2: "What goes in must come out—a separate way." For truly efficient operations, every entryway for goats should have a corresponding exit. It is almost hopeless to run an efficient operation (be it feeding, milking, clipping, or whatever), if you allow goats to enter and leave through the same door. The exception to this rule is if you have only one goat!

As I'm sure you've noticed, goats are extremely smart and can be greedy and aggressive. Since they learn quickly that they will be fed when they are milked, they all want to be first into the milkroom. This causes crowding at the door. Open a door to push out a milked goat, and three others will push to get in. This jamming is not only unpleasant but dangerous. Pregnant goats get squeezed, any goat can get ribs crushed, and legs (both goat and human) can get tangled causing further havoc.

Having an exit door allows your animals to act like the gentle folk they are meant to be, and the only jamming you have to contend with is at the entry door. Also, by always allowing only a set number of goats in at one time, your smarties will soon learn to count (or so it seems) and will eventually stand back and wait their turn.

#3: "Segregation works for goats." Having a way to segregate "to-be-milked" from "already-milked" goats is also a necessity when milking more than a few. Before we had holding pens (we did have separate entry/exits), after being milked, the most aggressive and greedy goats would run out the exit door, run through the barn, into the entryway and be next in line to be milked. Not only did this extend our chore time, but when people other than ourselves were doing chores, some goats came in to be milked three or more times before our relief milkers caught on!

#4: "More doors are always better." Having more than a single door out of the barn helps in moving animals about. Sometimes a particularly aggressive animal will stand in the doorway, making it impossible for you to move animals by her. If you have an alternative entry/exit, you'll be able to move animals more efficiently. And with more than one entry, you can always close one, while leaving the other open, allowing you to direct traffic flow as you need to. One door cuts down on your flexibility and your efficiency.

#5: While the saying "Idle hands do the Devil's work," may be a bit too strong here, you can gain efficiency if you keep both hands occupied or full whenever possible. This applies both to carrying and carrying out daily chores.

When you carry items, make every trip count. Look at what you have to carry and figure out how to move it using the fewest trips possible. To save strain on your back and limbs, use mechanical means (wheelbarrows, carts, etc.) whenever possible. Also, never leave a room or area empty-handed (or one-hand-full) without first looking around to see if there isn't something that needs to be transported to wherever you're going.

When you carry out various tasks around the goat yard, think in terms of using both hands or even other parts of your body to help you. For example, when bottle-feeding kids (a most inefficient means of feeding, by the way), never feed one when you can feed two or three at a time. As soon as kids are adept at drinking, you can feed two with a bottle in each hand and a third with a bottle held between your knees. (You can do this with calves, too, but holding the third bottle between your legs can be dangerous.)

Even when exercising goats at a fair, you can benefit from this rule. Try walking two goats at a time; you will halve the time it takes to exercise the herd. Use a mechanical aid, like a ring-side tie chain or other multi-ended lead, and you can exercise even more goats at one time. The only warning is to keep aggressive goats away from shyer ones, so you don't have chaos on the chain gang.

#6 is closely related to #5: "Never do something to one goat at a time when you can do the same to the whole herd."

The most obvious example here is kid feeding. When bottle feeding, you saw above that you can feed up to three at a time by using your hands and legs. However, to be truly efficient, you should consider a mechanical aid, something that allows you to feed a penful of kids all at once. These mechanical aids include gang feeders like the Caprine feeders or troughs that kids lap milk from. Both gang-feeders and troughs work well on kids that have been trained to use them, and the little bit of training it takes to get kids to "drink through a straw" on the gang-feeder or "lap like a dog" from the trough is worth the efficiency you gain. Any time you can feed ten or more kids in the time it takes to feed one, you're on your way to becoming an efficiency expert.

While milking a goat takes the same amount of time whether it's done singly or in groups, you'll gain efficiency if you milk in strings (bringing in more than one goat at a time). Remember that milking is not just drawing milk from an individual goat. There are numerous other parts to this operation: feeding grain, washing udders, drawing first milk into a strip cup, milking, stripping (for many who use milking machines), teat dipping, depositing milk in milk storage containers. It's far more efficient to perform these tasks with six goats at a time, going down a line, for example, than to do them individually with each goat.

Therefore, when possible, bring in the largest number of goats you can handle at one time and carry out the entire milking routine on all of them. Feed everyone grain, then wash udders, strip first milk, milk, teat dip, etc. You'll find that the old axiom "There is efficiency in numbers," is true even when it comes to milking goats. And your slow eaters will thank you, as they'll have time to finish their grain, while you are milking the rest of the string.

#7: "Look and analyze before you leap." For increased efficiency, break down complex operations to see where greatest efficiencies lie. I call this the "whole or part" decision. Before beginning a task, I think about whether it's more efficient to do the whole task at one time or to break it into sub-tasks.

The most obvious example is feeding. When all animals are grouped under one roof, it's probably most efficient to feed all animals hay at the same time, rather than doing the entire feeding operation (milk for kids, grain, and hay) separately for each pen of animals. If animals are housed under separate roofs, then the opposite is probably true. But generally, anytime you can carry out the same operation for the entire herd at once, you'll save time and energy over doing it individually for a separate group of animals.

I've found that this rule works especially well when clipping goats for shows. I can clip far more quickly if I do all the fine areas (legs, udder area, and head, for example) at one time on a whole show string. Then I go back and do bodies on all the girls. I don't know quite where the efficiencies come in, whether it's not having to deal with the same goat for an extended period or whether the repeated movements on these smaller areas makes me clip quicker, or whether I just don't get as bored with more goats moving through. All I know is that I get more goats clipped per day, doing it this way.

#8: Be a "clock watcher." Most of us live under the tyranny of the clock. We have numerous obligations that are governed by time (getting kids off to school, arriving at work on time, arranging to get does to a buck in time, etc.). Time or the clock may not be our friend.

However, efficiency experts live by the clock; it's what they measure most things by. How long it takes to do a certain task, how long a break between tasks, how many widgets get made in a set time, etc.

You can't improve the efficiency of your operation if you don't know how long you take to do things. Therefore, you must first understand just how long tasks take so you can decide whether the changes you make are really efficiencies or whether it's just the change—any change—that's making you feel better.

As my husband will testify, I can tell you how long virtually anything takes to accomplish. I can also tell you how much time I've saved on various efficiencies I've instituted. Armed with this knowledge, I can decide whether or not the efficiencies are truly efficiencies and worth keeping. If I cut two minutes off milking but increase my clean-up time, then that's not an improvement. If I cut time off milking but the animals are not happy with the changes, I can then decide whether the amount of time saved is worth the kind of disruption I've caused. Making the clock an ally is basic to true efficiency.

#9: Use the KISS principle whenever possible. KISS stands for "Keep it simple, Stupid." Of all the rules I've mentioned, this is perhaps the most difficult to put into practice.

For some reason, perhaps because many of us are perfectionists and want to do the best job possible no matter what or because we don't have confidence in our own ways of doing things, we tend to over-engineer, over-build, and over-do. The more complicated it is, the better it must be, we think. We look at comprehensive barn plans, milking parlors, and breeding programs. We think because they are "professional" looking, have lots of parts, and/or seem "scientific," they will be a better choice than the simple, workable, efficient barn, milkroom, or breeding plan that we've devised. Generally, the fewer steps, the fewer moving parts, the more natural it is or simpler it seems to work (and "it" here can mean almost anything), the easier it will be to work with, the smoother it will work, and the more efficient it will be for you and your goats. Go for "new and improved" or "state of the art" or "the best technology has to offer" only if "it" seems a sensible, efficient, workable choice.

The benefits of running an efficient operation are obvious. The less time it takes to do things means the more time you have to do other important things. If you count your labor when figuring the profitability of your operation, then you will realize a financial benefit from such efficiencies also.

Is the most efficient way always the best way to do things? Not always. There are times when we need to take the time to work with animals individually, and being an efficiency expert in your herd you can't forget that.

However, being your own efficiency expert and looking over your operation and figuring out how you can institute efficiencies will gain you both time and energy. You will also find areas where improvements will lead not only to increased efficiency but to the overall quality improvement of your operation and a better life for both you and your goats.

This article originally appeared in the Dairy Goat Journal *in 1996, and we reprint it here with our most sincere thanks to Dave Thompson, publisher.*

She's Got Individuality . . . Individuality

by Joan Vandergriff

As I was lying on the couch six weeks ago, arms wrapped hard against my stomach, trying to ward off another wave of gut-wrenching pain caused by a Latin American parasite, my husband waltzed by, looking fit and anxious for food. Food, I thought, that's probably how it all began. When the doctor had diagnosed the parasite (my dear husband kindly told all our friends I had worms!), he told me to think back over all the food and water I had consumed over the past two weeks to try to pinpoint exactly what had done me in.

Since my husband and I are pretty much joined-at-the-hip, especially when we're away from home, we brainstormed together and tried to figure out what, given the fact that I was miserably sick and he was gloriously healthy, I had consumed that he hadn't. Except for the copious amounts of the local Costa Rican rum that he had downed, our eating habits were virtually identical. While he would like to claim the medicinal qualities of our bottom-of-the barrel national rum drink, I knew better. The real difference between us was simply that we were two individuals, two different organic systems, two food-processing centers that reacted differently to the same stimulus. I have a tender stomach; his is cast iron!

Having plenty of couch-time to fill, my thoughts roamed from envy and anger at my spouse's gastric stability and strength, to my quests for individuality as a younger person, to instances of individuality that both dazzled and stunned me as I raised animals.

Nothing can be more frustrating than the failure to accept that all animals—goats and humans included—are individuals, reacting to similar situations and stimuli in widely divergent ways. Nowhere is individuality more evident than when it comes to breeding animals, and goats are no exceptions.

Catch a Rising Star

Individuality can be a blessing especially for those of us who are just beginners or breeders of stock that doesn't stand above the crowd. We hope and plan for improvement, but rarely is improvement anything but gradual. Most of the time, those of us with middle of the road (or middle of the show line) goats get used to the quality we can expect from the breedings we do.

And then one day, "Eureka!" One little head shines above the others. And others can mean mother, grandmother, sisters, littermates, it doesn't seem to matter. From the same gene pool of rather similar stock, a gem emerges. Just such happenings are the stuff that dreams are made of. Many a goat breeder, ready to sell out and raise Poodles, instead sees the glowing individual above all others and goes back to the herd, with renewed enthusiasm and strength.

No Such Thing as "More Unique"

Frustration comes when we don't remember that our great herd hope is simply an individual, and the dictionary reminds us that "individual" means "distinguished from others by special characteristics; of a unique or striking character." Remember that key word is "unique." We must also remember that the term "individual" is synonymous with "single," and therein lies the rub. The word "unique" means one of a kind, and one outstanding individual does not a whole herd make!

Most of us can look at the breedings we make and see certain qualities that we are trying to bring out (and some we don't want) reproduced throughout our herd. But sometimes something happens, and it (whatever "it" means) all comes together, quite differently from anything that has happened before, and we have our "individual," in all her (or his) unique splendor. We may make the same breeding that produced our gem time after time and never again reproduce it. That's what genetics is all about. Most of the time, it works the way it should. Sometimes it doesn't, and even fewer times, the result is exactly what we'd like to reproduce time after time . . . and probably can't.

But these outstanding (unique) individuals serve their purpose and do advance our efforts in their own ways. Obviously, seeing that world-beater daily out in our goat pen is a joy to those of us used to seeing its more garden-variety herdmates. This one goat assumes great power in our lives, injecting us with energy and enthusiasm to improve the rest of our herd. If we show this animal, and it does well, we gain added respect and interest in our stock. And if we use our brains and do a little research, we can try to figure out just what it is that makes our star a true star, and with a little more luck, we can develop a breeding strategy that might increase the likelihood of reproducing another standout sometime in the future.

Spreading Uniqueness in Our Herd

We need to figure out just what is it that gives our standout that individuality. Is she taller, better in the front, higher and tighter in all udder attachments, a more persistent milker? It's important to decide whether she is just better than the rest of her kin in a single dominant quality or a number of small qualities that combine in a complete package. To do this, compare your star with her siblings and other relatives. Note every quality that she has that the others don't have. Conversely, look at the qualities of her relatives to see if they have some positive qualities missing in your star. After all, we need to recognize the strong traits already residing in our animals, and we don't want to lose those in the process of improving other qualities.

Of Princes and Frogs

Now comes the research part. As with the princess who had to kiss a lot of frogs before she found her prince, you won't have to kiss frogs, but you'll need to look at lots and lots of animals either in person or through the mail, magazines, or Internet in order to find the would-be-prince/princess for your herd.

Begin by looking at pictures (on web sites or in magazines) or go to shows. Find a herd that seems to have the outstanding qualities of your single standout that are lacking in your other animals.

Talk with or write to breeders, and ask questions. What did the foundation stock look like? What kind of animals did the breeder use to get the characteristics you're seeing? What kinds of specific qualities (both good and bad) are being passed on, and what percentage of offspring exhibit the qualities you are looking for. What happens when the line is bred to totally unrelated animals (like yours); what happens with linebreeding or inbreeding?

Get as much information as you can, and try to see as many examples of the stock as you can in as many (preferably unrelated) herds as you can. What you end up with after all this research is a rough idea of how this breeder's stock may (and here the word "may" is most important) work with your herd.

Remember the Tortoise and the Hare

If you decide to invest in an animal from someone else's herd, remember that you can't expect miracles. What you are looking for is either specific improvement of the individual trait you are focusing on or general, across-the-board improvement—all of this without losing positive traits inherent in your herd, or without introducing new negative traits.

Don't expect that every offspring will shine like the outstanding individual that began all of this. What you're looking for is consistent movement over time toward the qualities of your star.

Nature or Nurture?

There are many other instances where we need to recognize individuality within our herd. These may not be as herd-shaking as the standout goat, but they can be frustrating.

Just as I was felled by a parasite that left my husband unscathed, so can any number of ailments do the same in your herd. Why do certain individuals always get snotty noses on cool nights or at shows? Why does one kid always have some kind of skin rash or ringworm while her pen-mates are clean? Other aspects relate to breeding individuality. How come one doe is easy to A.I. and her twin has never been A.I.ed successfully? One always breeds on the first cycle, her mother never until late in the season?

You can extend this list endlessly with examples from your herd. The gist of all this is that even though you may have animals that are closely related genetically, animals that have been brought up in exactly the same environment, animals that are alike in almost every way, they may differ from the rest in some particular (and important) way.

The trick is to recognize each animal's individuality and deal with it to the benefit of the animal itself and the herd as a whole. I have lost animals simply because I failed to recognize the importance of an individual's specific trait or weakness. That failure hurts not only because of the senseless loss but because of how easy it would have been to head off the problem.

Recognition of each animal's individuality and then the constant observation of these animals are the keys to success here. It doesn't take long for us to realize that our favorite, Dolly, has a touchy stomach like me. Feed her too much grain and she'll have problems. A little dust or mold in the hay and she'll let you know almost immediately with the runs. Knowing her potential for problems of this type makes us take care about the amount of grain we feed her and inspect the hay we feed her carefully.

The animal that always gets sick at shows, probably deserves to stay home or should be pampered continually through her stay. A pen out of the draft and away from the hubbub of fairgoers and other people's animals, a t-shirt or goat coat at night or when it's cold, a regimen of an anti-stress agent such as Probiotic Powder or Nutri-Drench, the freshest water and the most appetizing hay, all work to head off this goat's particular weakness. Recognize the "Typhoid Dollys" in your herd, watch them daily, and treat them quickly the moment you suspect that something might just be amiss.

Don't Kid Yourself

A kid's uniqueness is often a bit harder to deal with. First, most kids seem to be eating, frolicking, and sleeping machines from birth through the first month. Later they develop true, lasting personalities and individual qualities that we can look forward to (or cringe about) for their lifetime. Here, the person who does the feeding and who spends the most time with the kids must be the one to recognize that "slow" kid that just can't quite suck milk up a straw on a gang-feeder like the rest of them (try a bottle or a pan), or the "wimp" who lacks the aggressiveness to get in there and eat grain or hay (take it out and feed it separately until it gets a little older) or gets beat up by those a few days older (segregate it with a few other less aggressive kids).

Fail to recognize these individual traits and you'll have a kid that fails to thrive or actually gets hurt if left with other babies.

It's All in the Breeding

Breeding quirks are a bit easier to deal with. Since most of us keep careful records of our does' heat cycles and breedings, at the end of a single season, with good records, we have a good tool to help us recognize those animals that have "special needs."

We should be able to know when a heat cycle began and ended each time, how long a heat cycle lasted, when in the cycle she was most interested in the buck, when she was bred and whether that breeding "took," what part of the cycle she seemed most fertile, etc. Most of our goats have normal heat cycles, breed and settle when we want them to, and have a normal gestation period. By studying your breeding and kidding records, you will quickly be able to identify does that are difficult breeders; those with irregular heat cycles; those that fail to breed even when in obvious standing heat; those who carry their kids for a longer or shorter than normal gestation period.

Each piece of information we get on individual does means, when it comes to breeding, the greater the likelihood of our getting our does bred when we want them bred (not when she wants to!). Knowing the idiosyncrasies of our animals allows us to minimize the frustration that comes with animals that fail to breed easily. Since we know a doe's specific problems, we can take measures to insure that we get the job done.

A Manly Man

Breeding idiosyncrasies can work both ways. I once had a buck who was shy. Early in the breeding season, I put does in heat in the pen with him and allowed them to mate as they would, leaving them together for a day or so.

When none of the does settled and the artificial insemination collector declared the buck fully fertile, I had to look to some personal quality that was interfering with his ability to breed does.

On the next occasion, I put the doe in and watched. What I had was a "wuss." Having been brought up as a single buck, alone in his pen, with only people for friends, my "manly man" was scared stiff of any doe, especially an aggressively amorous one, who wanted to get up close and personal. The answer was simple. I brought him out and held onto the doe, allowing him to get used to her without her bossing him around. Soon he figured out what his mission was, and after a few times of being the boss, he was much more anxious to breed does who were put in the pen with him.

You're Either With Us or Out the Door

Not all such behavioral idiosyncrasies can be overcome so easily. Nothing brings out the potential for "goat abuse" faster than goats with annoying habits. The trick is to identify these miscreants fast, before the behavior gets thoroughly ingrained. If a goat distinguishes herself by not allowing herself to be caught, break that habit quickly. Goats, being reasonably bright, eventually learn that not sneaking into the milkroom repeatedly makes our life and thus their life happier. For the "right-minded goat," we will call them by their own name, often with endearments, in a kind, gentle tone of voice rather than the alternative. In my household, those dears that don't learn to overcome offensive personal traits find themselves out the door, fast, to someone with more patience than I, to the auction, or to the butchershop.

I remember from my youth the song, "Personality, personality, —she's got personality." Personality then was a good thing. It denoted spunk and character and individuality. Today, we don't seem to embrace individuality quite as much as we did years ago. It's more comfortable if we don't stand out. Actually, most of us are ambivalent. We may envy those among us who are a little bit more daring than we are, those who live their lives a little outside the mainstream, those who make life choices that we'd like to make if we didn't have all those responsibilities that tie us to a more regimented life. However, while we may envy their spirit, we're more comfortable with the life we're living.

And when we have to deal with people with "personality" or individuality daily, it's often frustrating since we don't see the world from the same point of view. The same kinds of frustrations arise when we deal with our goats, be they our shining stars, our Typhoid Dollys, our reluctant bucks, or our exuberant gate-crashers. Usually, recognizing their individual traits and working with them will lessen our frustrations and bring them back into our mainstream. The others, well, they'll probably be more comfortable with other goats that also wear nose rings.

Hoof Trimming: An Overview

Hoof trimming is one of the most neglected chores of all. And if we really think about it, allowing our goats to walk around on overgrown hooves can not only be uncomfortable for the goat but potentially harmful.

Just remember how your feet feel after a day walking around in shoes that do not fit right. If you persist in wearing those too tight or too loose shoes, you may find that you use your hips to adjust your gait to make walking more comfortable, knocking your torso out of normal alignment. Goats do the same thing.

Also, once the overgrown hoof begins to turn over, the flap it forms on the hoof bottom becomes a natural cranny for dirt and bacteria to hide. Eventually, especially in damper climates, hoof rot fungus can take hold, and once in place, it can be difficult to eliminate. Severe cases of hoof rot can make walking almost impossible.

To understand the ideal form you are aiming for when you trim hooves, look at a newborn kid's foot. You will see the proper shape you should be trimming for. Rehabilitating badly overgrown hooves will take more than one trimming. For those truly horrible hooves, take off a little bit at a time. Remember that each time you trim off hoof, you are reshaping the platform the goat will walk on, and the goat must get used to a new gait your trimming results in.

No one can say how often you should trim the hooves on your goats. Just like humans whose hair or fingernails grow at varied rates, some goats will grow their hooves more quickly than others. As with other aspects of raising goats, it is important to know each of your goats and their needs, and trim their hooves as needed.

Old-time breeders used a sharp knife such as a pocket knife or utility knife to trim hooves. Today, we are lucky to have wonderful hoof trimmers that make the job a whole lot easier and safer. Our choice for trimmers is the orange-handled one with teflon-coated blades. Its pointed ends help you clean out dirt embedded in hooves, and the narrow blades make for accurate cutting. Finally, these trimmers are small and lightweight enough for older children and adults with small hands to use comfortably. A warning, however: these hoof trimmers are incredibly sharp and stay that way. We've cut ourselves often, so use them with care.

Anatomy of a Hoof

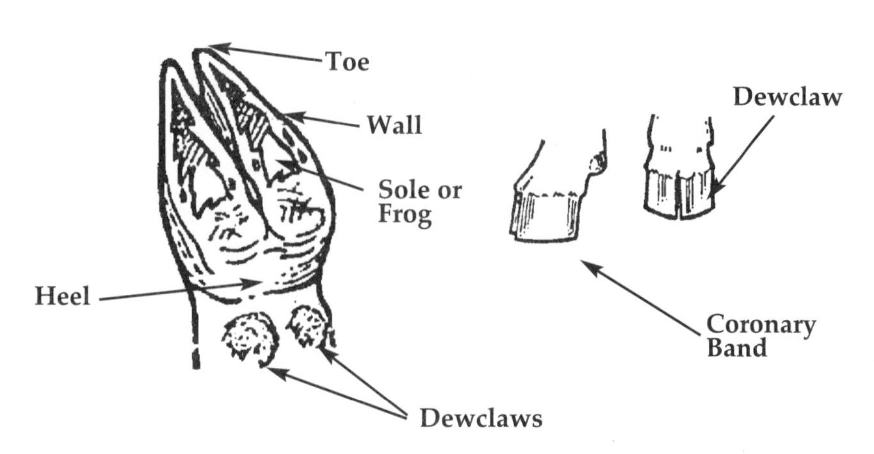

When a goat walks, it should be walking on the walls of the hoof, not the sole. What this means is that the sole should not be rounded but flat, and the hoof should be an even height front to back (the coronary band should be an equal distance from the hoof bottom from front to back).

You can not do a good job trimming hooves unless the goat is tied securely or immobilized in some way. The easiest way to do this is to secure your goat in the milk stand or on a clipping stand if you have one. If you choose to tie your goat, try to do so next to a wall, so the goat can not move away from you as you hold up its leg.

You can start with a front or rear foot, it makes no difference.

There are two ways to stand when you trim front feet on goats that are tied. You can either kneel in front of the goat, facing the rear and pick up the foot (this works primarily on calm goats since you have no real leverage for holding the goat still). Or you can stand beside the goat, lean over its shoulder

and pick up the opposite foot, holding it tight against the goat's body. This position gives you a firm hold on the goat and the foot. With goats on a milk stand, you can stand to the side of the goat and simply pick up the foot. (Most goats will tend to stand still on the stand as they do not want to risk falling off.)

To trim rear feet, stand over the goat, facing the rear, and pick up a foot and bring it back and slightly up so you have some leverage and can work on it. You can use the same stance for tied goats or goats on a milk stand. Obviously children will have trouble standing over grown goats, so you will want to help them trim the feet of larger animals.

*The above three illustrations are reprinted with permission from Hoegger Supply.

A Step-by-Step Guide to Hoof Trimming

1) Begin by cleaning any accumulated dirt out from the front area of the hoof bottom using the point of your trimmer.
2) Next, you may find that the walls are long. They should be cut to the same level as the bottom or the sole of the hoof.
3) The sole or frog should be flat, not rounded, so you may need to level it. Do this by cutting a little bit at a time. Do not try just one cut to reshape the sole. Stop when the hoof bottom looks a little pink.
4) If necessary, trim the heel so that it is level with the sole and wall.

1) Clean out dirt

2) Cut the wall down even with the hoof bottom

3) Cut the sole or frog so that it is flat

4) Trim heel to match wall and sole level

Before

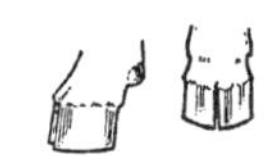

After

Hoof Trimming: Hints

"Instead of using trimmers, you can use a hoof plane or rasp to flatten the sole and finish the hoof. Children can do a good job with a hoof plane if they are taught how to use it properly from the beginning and if they trim hooves on a regular basis. Then, most of their work will be taking off a little excessive growth on the hoof bottom."

"Hooves are much easier to trim when they are wet. Try trimming hooves after the goat has been walking in snow, mud, or morning dew."

"Hoof trimming is a chore most of us don't do often enough, but I've found a way to get it done without it killing me. And it will work no matter if your herd is 10 or 100. Every time I milk I trim the hooves of a set number of goats, that's just part of my milking routine. And I continue to do this, without fail, until the whole herd has had its hooves trimmed. That way I get the job done in a reasonable amount of time, and it's less back-breaking than doing them all at once."

"Our goats kept getting knocked down in the shows for bad feet. We began a routine of hoof trimming, and we found our goats' bad feet had been caused by bad management, not bad genetics!"

"Most goats don't like their hooves trimmed. You aren't hurting them; they just don't like standing on three legs."

"Try not to trim your goat's hooves at a show. Trimming should be done a few days before the show to give your goat a chance to get used to how the foot meets the ground. It's always possible to trim too close, and there's nothing worse than having your best doe limp around the ring just because you've trimmed her hooves too close."

Milking: An Overview

When you decide to raise dairy goats, if you are serious about this undertaking, one of the decisions you are making (whether you know it or not) is to milk your does, morning and night, at least ten months a year.

Like other animals, when a doe kids, she usually comes into a flush of milk, meant to feed her kids. Many does soon have more milk than their kids can consume, and breeders, even if they allow their kids to nurse their moms, may milk out the extra milk regularly. If you choose to raise your kids by hand (not allowing them to nurse their dams), you will need to milk your does twice a day. Such milking will usually encourage does to increase their production.

Most does' milk production will follow the classic bell curve, beginning often with a modest amount of milk and increasing incrementally over a period of three or four months, peaking and holding for a period of time, and then tapering off.

Typically, a doe will freshen in the spring. You will then milk her for seven months, breed her, and then dry her up when she has been milking for ten months. (In some parts of the country, depending upon length of daylight and climate, among other things, does may freshen off season.) This should give your doe the last two months of her pregnancy to build herself up without the stress of milking.

Some does will dry themselves up. Others with strong wills to milk may be impossible to dry up. To encourage a doe to dry up, you may try milking her once a day until her production is so low that she doesn't need milking at all. There are a few persistent milkers that will never dry up.

The most basic milking set-up can be nothing more than a tie chain to immobilize the doe while milking and a bucket. However, if you plan on milking more than one or two does, you should consider using a milk stand. It will save your back. Also, if you plan to use the milk, you will need a strainer with filters and a container for storing milk. Finally, for your does' health, you should consider using some mastitis screening device and a teat dip.

How to Milk and Handle Milk

Milk your does in a clean place, away from dirty bedding and manure. Milking will be easier if your does stand on a raised milking stand. Your milk will be cleaner and your work easier.

For your milk to be clean, your does must be clean also. Clip long hair from udder, flanks, and tails because long hair catches dirt which can fall into the milk. (If you visit commercial dairies, you may see milking does with this type of abbreviated haircut, meant to keep milk cleaner.) Brush dirt from your does before you milk. Next, wash udders, and then dry each udder with a clean paper towel. Use a separate towel for each doe.

Before you begin the actual milking, draw and discard the first stream of milk from each teat. Any dirt at the teat opening will come out with this milk. If you use a strip cup to catch this milk, you can check for signs of mastitis at the same time.

Protect fresh milk from flies, dust, and other dirt. Putting a cover on the milk pail between each doe you milk will help. Next, strain the milk through a strainer with a clean filter disk as soon as possible. And never re-use those paper filters!

For the best quality milk, cool your milk quickly. For the fastest chilling, set the container of milk in a tub of water and ice, and stir frequently. You want your milk chilled to 40°F quickly, and your refrigerator can not do that. Finally, do not mix fresh, warm milk with chilled milk.

Milking Equipment

The rules that commercial dairies must follow are meant to ensure that the milk they produce is the cleanest, most healthful possible. Rules about milking equipment focus on the materials it is made from. Some materials are too porous, allowing bacteria and dirt to build up. Cracked or pitted equipment will also harbor bacteria and dirt.

Stainless steel and glass are approved for equipment in Grade A dairies, because equipment made from these materials is easier to keep clean and does not tend to harbor bacteria or dirt as readily. In a non-commercial setting, which includes most goat owners' operations, you do not need to use stainless steel pails and containers. However, doing so will help improve the quality of the milk you produce.

You can use tinplate pails and containers if the inside seams are smoothly finished, not worn and rusty. If you use aluminum pails, clean them with gentle, non-abrasive cleansers as they tend to scratch and pit easily. Use enamelware only if there are no chips or gaps in the finish.

Most plastic is too porous to be sanitary. You can not clean it thoroughly, and plastic can harbor bacteria, causing off-flavored milk. There is, however, a special food-grade plastic used for some dairy equipment, cheese molds, and milk jugs.

All of the equipment you use to handle milk should be seamless, or if it is pieced, the inside seams should be smoothly finished so no dirt can collect.

The basic equipment you will need if your budget is tight includes the following: a **milking pail** (a stainless steel mixing bowl will work), a **milk filter** with paper filter disks to strain the milk—a household mini-strainer will do if you are filtering milk from one or two does, a **box of milk filters** (use a new, clean filter each time you strain milk), a **pail or jars for storing your milk.** Wide-mouthed canning jars or gallon glass jars work well here. In the past you could pick up free gallon glass jars from fast food restaurants. Unfortunately, today most restaurants use plastic gallon jars, which are not a good choice for storing milk (it is almost impossible to get that pickle smell out of the plastic, not to mention the porous quality of the plastic itself).

Cleaning Milking Equipment

To have clean, good quality milk, everything the milk touches must be clean. While this may seem easy to accomplish, you should first consider that the environment most people milk in is not the most sanitary. Therefore, you begin with a basic problem that your milk handling and cleaning of equipment must overcome. Keeping your milking area scrupulously clean and free from flies and other insects will help. However, nothing substitutes for clean equipment.

For best results, immediately after you finish milking and storing your milk, rinse all of your equipment with cold water and then with warm water which helps rinse off the milk fat. Scrub all equipment thoroughly with detergent and a stiff brush. There are dairy detergents made specifically for this purpose, and they will probably work better than kitchen cleanser. Next, rinse your equipment well with hot water. Finally, for sanitizing, the manufacturers of Clorox recommend that you add one ounce (two tablespoons) of Clorox bleach to two gallons of water and rinse your equipment. Store equipment away from flies and dirt.

In commercial dairies, equipment is rinsed again with a chlorine sanitizer just before it is used. If you want to sanitize your equipment again before milking, use the same sanitizing solution (one ounce of Clorox bleach to two gallons of water). Rinse all equipment in this solution immediately before use and allow it to drain thoroughly. Jars and bottles you use to store milk in should be immersed in this solution for five minutes, then rinsed and drained before filling.

Stainless steel equipment that is left to air dry may have water spots or other stains. You may want to dry your stainless steel equipment with a soft dry cloth.

Milking by Machine

There is a lot of misinformation and old wives' tales about machine milking.

If you have just a few goats, you probably will not save much time, but milking even a few goats by machine is easier physically than hand milking. With more goats it is both more efficient and less exhausting.

Some people are afraid to use a milking machine because they think it can cause mastitis. In general, machine milking does not cause mastitis; careless milking and improper use of the equipment does.

There are a few basic rules to follow when milking by machine.

First, always check the vacuum prior to attaching the inflations to a doe's teats. Too high a vacuum level can lead to mastitis because of the continued increased "pull" on the teats and irritation to the udder. Setting the vacuum level is a balancing act between efficient milking and good udder health. The manufacturer of your machine should know what vacuum is best for your set-up.

Be sure not to over-milk. Over-milking is probably the most common cause for mastitis when milking by machine. To guard against over-milking, monitor the milking whenever the machine's inflations are attached to the doe and the machine is running. The moment you see milk flow lessen or stop, remove the inflations.

For most novice machine milkers, plastic shells and silicone inflations are a good choice because they let you watch the milk flow. The moment the milk flow stops, you can see it and shut off that inflation. Also be sure to put some sort of clamp or shut-off on each milk line, so you can shut off each inflation individually.

Use a strip cup to sample milk from each doe prior to milking. We recommend that you hand strip does after machine milking. Some people do not do this, but it seems the safer course. Leaving milk in the udder can cause disease and decrease production. Use a good teat dip when you are finished.

People may have trouble with milking machines that are jerry-rigged for goats. For best results, choose new, quality equipment from people who know both the equipment and goats. Be sure you have someone to talk with if you run into problems.

If you follow these simple guidelines, you will find milking by machine can be good for your goats as well as for you.

Milking Machines

While milking machines are not terribly complicated, you need to know a few simple principles. The motor (which can be gas powered but is most likely powered by electricity) provides the power that runs the pump. The pump that milking machines use is a vacuum pump, and it is used to develop a vacuum in a tank. The tank has connections to the milk bucket assembly. The milk bucket assembly, which is composed of bucket and lid, milk lines, vacuum hoses, shells, and inflations, attaches through the inflations to the goat's teats.

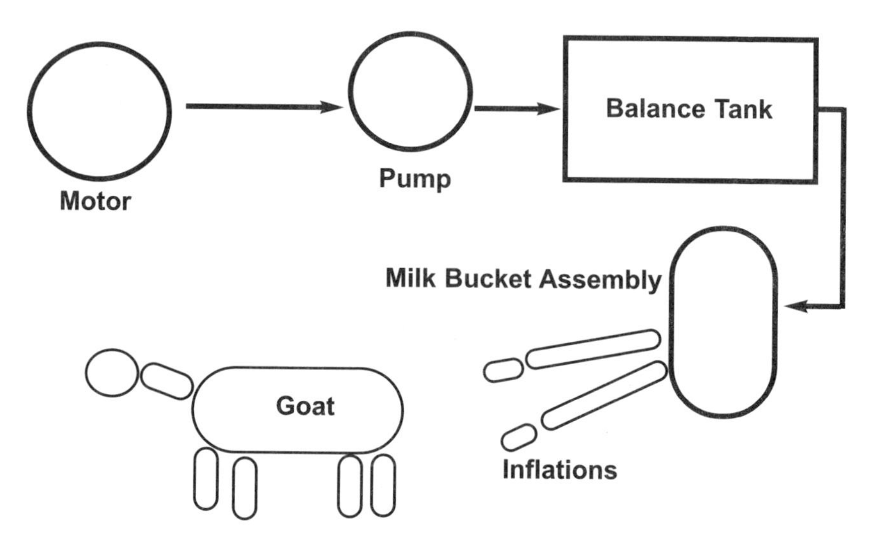

A regulator holds the vacuum constant when the vacuum is present, and a pulsator regulates the timing of the actual "milking action." As the vacuum comes on, it actually sucks out the milk. The milk flows through the milk lines into the milk bucket. When the vacuum goes off, the inflation relaxes, allowing milk from the udder to fill the teat once again. And the process starts all over again.

On the next page are two different types of milking machines. The first is a portable, "cart" model. it shows three different bucket options. Cart models can be used as an everyday milking machine, and you can buy them set up to milk one or more goats. Since they are fairly lightweight and come with wheels, they are handy to take to shows and can be used to milk does (or cows) whose milk needs to be kept separate from the main herd's milk. The second machine would be appropriate for a more permanent installation. This model is set up to milk two goats. These "permanent" systems can be set up to milk pretty much as many goats as you want. The limiting factors are the size of the motor, pump, balance tank, and ability of the human milker and milking parlor to handle a large number of goats. No one can tell you how many goats it takes to justify using a milking machine rather than hand milking. That's a matter of personal choice and circumstances. For some people, milking two goats by hand is just too much, while for others milking twenty or thirty by hand is an easy day. Physical considerations also play a part. Some people are just not physically capable of day-in, day-out hand milking.

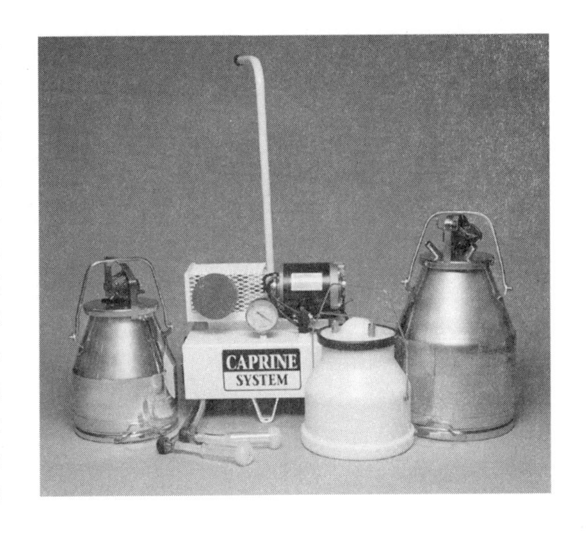

All things being equal, if you are milking only a very few goats, it's a matter of weighing the additional time and physical effort it takes to milk by hand against the time it takes to set up, disassemble, and clean equipment when milking by machine.

Milking systems can be set up to milk just about as many goats as you want at one time. The limiting factors are the size of your milking facility and the number of goats the human handling the milking can supervise properly at one time.

Another factor to consider is whether you want a portable or permanent model. If you plan to milk your goats in more than one place or want to milk by machine at a show or facility other than home, then you should consider a portable, cart model. For those who want to set up a permanent milking parlor facility, then you may decide on one of the larger models that feature a separate tank. These models allow you to place the motor, pump, and tank outside the milk room, which cuts down on the noise levels. Generally, well-designed portable models are not noisy when running. Some goat owners set up a permanent milking facility and also have a portable system to use for milking goats whose milk must be segregated or to take to shows.

Finally, if you plan to market your milk, then you have another decision. Since milk must be kept scrupulously clean in order to sell it, the less the milk is exposed to air, the cleaner it will be. Many commercial producers install what is called a "pipeline" system. Here, milk from each goat milked goes from the goat through a "hard" or "pipeline" directly to a refrigerated bulk tank. Milk is not exposed to air on the way to the bulk tank. Producers who choose to use a bucket system must take the milk from the milk bucket reservoir and pour it into a bulk or other refrigerated container. Milk is exposed to air as its being transferred. In many states commercial producers may not use a bucket milking system

Milking Machine Components

Motor: The motor (which is usually electric, but there are some gas-powered motors) provides power to the pump.

Pump: The pump is the "heart" of the milking system.

Tank or Balance Tank: The balance tank is the container that allows the pump to develop a vacuum.

Regulator: The regulator, attached to the balance tank, holds the vacuum at a specific level and smooths out any fluctuations in vacuum level.

Pulsator: The pulsator controls the milking, switching vacuum on and off to the inflations. Pulsators are set at a 60:40 ratio.

Milking Bucket Assembly: The milking bucket, with or without pulsator, milk and vacuum lines, shells, inflations, and claws.

Claws: Claws are a holdover from traditional cow systems. They act as a reservoir for milk. Most new systems use the a goat claw, a plastic claw with an automatic shutoff.

Inflations and Shells: Inflations are the liners inside the shells that attach to the goat's teats. They may be made of silicon or rubber.

Milk Lines: Nowadays, milk lines are clear and allow you to see milk flow all along the line.

Weighing Milk

The right way to measure the milk output from your does is to use a scale and weigh the milk produced each milking. Weighing milk is a lot faster and easier that measuring it, and a dairy scale is a valuable piece of dairy equipment. By weighing milk regularly, not only will you get to understand the cyclical nature of your does' production, but you can also spot problems early by noticing changes in milk quantity.

To help you weigh milk, it is good to know that a pint of milk weighs about one pound. A gallon of goat milk weighs about 8.6 pounds, but the pound-per-pint rule is easy to remember. Milk high in butterfat weighs a bit less.

If you want good information about your does' production, for every doe you milk, weigh milk and write down the weight once a week. By doing this, you will notice any drop (or increase) in production. A drop in production can alert you to health problems or a doe coming into heat. You can see the results of changes in your feeding program (such as a new batch of hay or added minerals) just by looking at your "in-house" production records. If you have production information for every doe in your herd, you can use it to help you plan breedings.

Your state agricultural extension office or university may offer official milk production testing and other programs that may be helpful to you. Once a month you weigh milk officially, take a milk sample for analysis, fill in the official records sheet, and get back computerized records that can be extremely helpful to you as a breeder. Contact your county extension agent or breed association for details, or call your local goat club. These sources should have information about programs available in your area.

If you do not want to be on official test but want to estimate your doe's production for a month, take an average day's production and multiply it by the number of days in that month. This will give you an idea of your doe's monthly output.

Average Milk Production

Some exceptional does produce seven or eight quarts of milk a day when they are fresh. Other does have nice personalities or spots, but the best they can milk is less than three quarts a day in peak production. Obviously, milk production varies.

On the average, you will get three-quarters of a gallon of milk a day per doe. The most recent (2008) actual average milk production (according to the USDA) for Alpines, Experimentals, LaManchas, Mixed Breeds, Nubians, Oberhaslis, Saanens, and Toggenburgs on official DHI production test was approximately 1739 lbs. of milk, 64 lbs. of fat (3.66%), 54 lbs. of protein (3.04%) for ten months.

A good mature doe (three to four years old) will give three to five quarts a day when fresh and should still be producing two to three quarts a day seven months later when it is time to breed her again. Some breeds average less milk, but their milk may be higher in butterfat.

Young does usually produce less. Some breeders are pleased with the first freshener that hits three quarts a day at her peak. A first freshener that milks too much may break down and not be able to continue her production in the years to come. If a young doe peaks at only 5 lbs., you can give her a chance, and she may continue to increase her production each year until she is mature. The ideal doe milks consistently over a number of years.

When a doe is in heat (and also after she is bred) her production may drop off sharply. A drop in production may also signal illness.

Production depends on genetics, condition, management, even weather. Remember: it also takes quality grain, good hay, and lots of clean fresh water for a doe to milk well.

Average Milk Production

The 2008 USDA records (reprinted from the USDA Animal Research Service website at http://www.aipl.arsusda.gov) report the following average milk production (in pounds), butterfat (in pounds and by percent), and protein production (in pounds and by percent) by breed:

	Milk	Butterfat		Protein	
Alpine	2091	69	3.3%	61	2.92%
LaMancha	1866	71	3.8%	56	3.0%
Nubian	1424	65	4.6%	52	3.7%
Oberhasli	1683	60	3.6%	53	3.1%
Saanen	1922	64	3.3%	56	2.9%
Toggenburg	1694	54	3.2%	46	2.7%

The 2008 records show 386 herds in the United States with officially reported records. The top ten states in terms of number of herds on test are:

California	36	Maine	16
Oregon	33	Ohio	15
Washington	27	Michigan	15
Wisconsin	20	Pennsylvania	12
Minnesota	20	Indiana	11

These figures, while reporting only those herds with official tests, will give you a good idea of the areas of the United States where goat populations are the largest and participation in production programs the most popular.

It is interesting to compare the above figures with those for 1997 (the first edition of this book). In most cases, states have shown a decrease of 1/3 or more herds on test. Where Wisconsin had 53 herds on test, last year saw only 20. Likewise, California, while 3rd on this list with 47 herds in 1997, in 2008, it led all states with 36 herds on test.

Off-Flavored Milk

Fresh, clean goat milk should have an excellent flavor with no special taste at all. Most people find it tastes "just like regular raw cow milk," perhaps a bit sweeter, and actually better-flavored than store-bought milk.

But not all goat milk tastes good.

Some off-flavors are caused by dirty equipment or not cooling the milk properly. Off-flavors can come from what the goat eats—strong-flavored weeds, etc. If your does go out to pasture, try bringing the does in from the pasture three hours before milking and feeding grain and hay after milking.

A wormy doe may give odd-tasting milk. A heavy-producing fresh doe may have slight ketosis, and this may flavor her milk. Any sudden feed change such as a move to a new home, introduction of new goats into the herd, etc., can upset a doe's metabolism and cause a flavor change in the milk.

Subclinical mastitis can cause the milk to taste odd.

Some does simply do not produce good tasting milk. The problem can be genetic and run in families. Keep milk from these does separate from "house" milk.

Raw goat milk will eventually turn "goaty" and off-flavored due to capruic acid. The rancid aftertaste may be noticeable within a couple of days with some goat milk. In other cases, the milk may stay fresh-tasting a week or more. If very fresh milk is pasteurized, the enzyme that produces that off-flavor is destroyed, and the flavor will stay good for a long time. To pasteurize milk, simply heat the fresh milk to 165°F, and then cool it. This may take care of the off-flavor problem.

Hints for Eliminating Off-Flavored Milk

"We had problems with off-flavored milk, and after trying a number of folk remedies, we decided that we had to attack the problem on more than one front. It worked. Here's how you can help overcome 'goaty' milk. Make sure your milk is scrupulously clean and chill it quickly. All dairy equipment must be 'really' clean. A dairy rinse can help here. Brush goats and wash and dry udders to make sure nothing falls into the milk. Filter the milk and chill it immediately. You want it to get to 40°F quickly, and a refrigerator won't do that.

An ice-water bath does a fine job. Test whether the milk from all goats is 'goaty,' or only one. If all milk is bad, look for a feed or cleanliness problem. If one goat is the culprit, don't use that milk for fluid consumption, cooking, or cheese. If "goatiness" in milk seems to run in a family, consider a change in breeding programs. 'Goaty' milk may result from a combination of factors, and besides if you do everything we suggest, not only will your milk be good tasting, but the overall quality of all your milk will be better."

"We have eliminated bad-tasting milk from does in our herd with the use of Vitamin E. First, we made sure that every goat was properly wormed. Then if her milk was still strong flavored, we gave her 2,000 I.U. Vitamin E each day for five days. This has improved the taste of milk from does whose milk was truly undrinkable."

"We cured several cases of off-flavored milk with garlic! We give the doe two tablespoons of baking soda and one clove of garlic minced (we use the minced garlic in jars) on the doe's feed every day for two or three weeks. I don't know why it works, but it does. (Ed.—Maybe the strong "goaty" taste is masked by the taste of garlic in the milk? No reports from other breeders on this remedy.)

Milking: Hints

"I find teat tape is easier and quicker to remove if you turn down one corner just a little on the last piece you put on."

"We use our 9-quart milking pail not only for milking but for making cheese, feeding calves milk, and for our veterinarian to wash her hands in when she visits our farm. After ten years, it's still bright and shiny. Love that stainless steel!"

Housing, Feeders, and Fencing

We begin this section with a simple plan for housing goats.

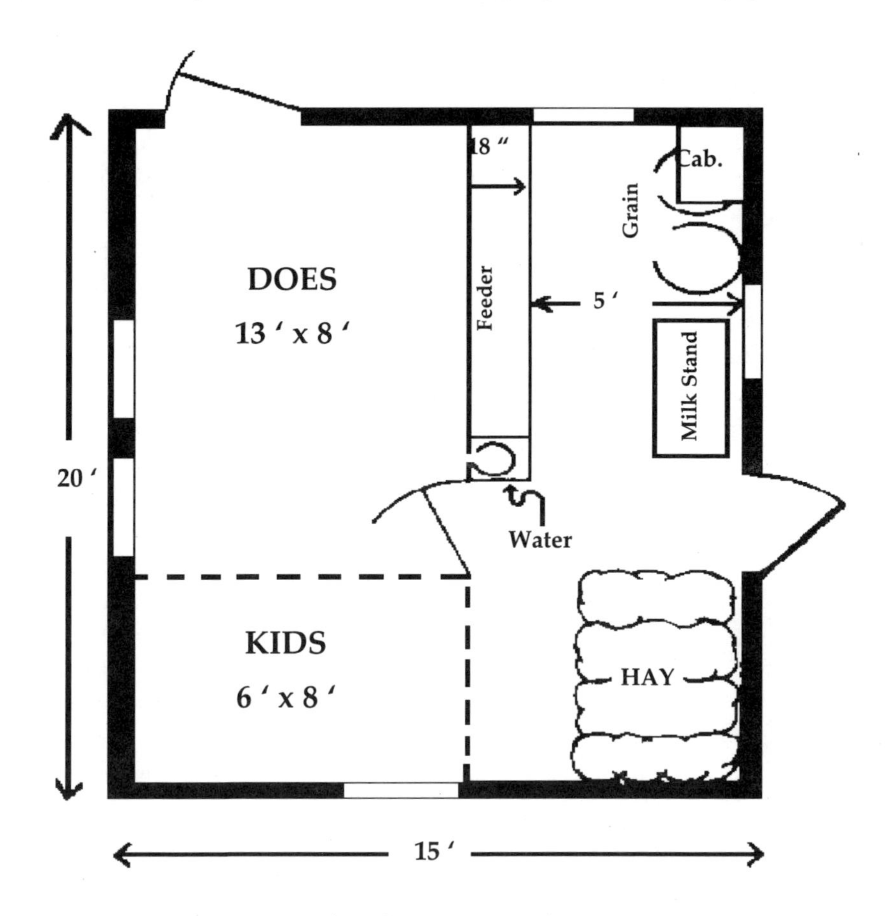

This plan is not to scale.

This plan is intended to house a few goats in a small (15' x 20') space. A good rule to follow is to allow 25 square feet for every goat. Therefore, the 13' x 8' doe loafing area (the area where goats spend most of their time) will serve 4 goats with room to spare. The feeder (for both hay and grain) has head openings spaced 18" to 24" apart. This allows more than enough eating space for four does. With this design, goats also reach through a head opening to drink water.

Since the water tub sits on a shelf outside the pen, water will stay much cleaner than water left in buckets inside a pen. Notice that with this design you do not need to enter the pen in order to feed goats or change their water.

In this plan, you also have a separate pen for kids. This is important when kids are young, as they are often intimidated by grown does and can be injured easily if housed with older, larger animals. You can use wall-style feeders for hay and as in your doe pen, place the water in a bucket outside the pen.

Since goat kids (like their human counterparts) like to play with their food and water, place the feeders high enough to keep kid feet out, and forcing kids to put their heads through some kind of opening to drink from water buckets outside the pen will help keep the water clean.

In the work space, there is room for a milk stand, two 50-gallon metal or plastic containers for grain, and a wall cabinet for supplies and equipment. For flexibility, use movable panels for the kids' area. At certain times of the year, you can remove the panel between the kid and doe pen and use the whole space for your does, or you can remove the panel separating the kid pen and the work area for additional hay storage. Half-length hog panels are excellent for this kind of flexible space. You can attach them to the walls using eye-bolts and snaps and attach them together at the corner with spring hooks or cable clamps.

The most important thing to remember in designing space for goats is to allow for flexibility, efficiency, and sufficient area for the goats you plan to house. The plan above is fine for 4-5 does plus kids. But you can expand it to house larger numbers and use the ideas above to make your goat housing more flexible and efficient.

No-Waste Feeders

No-waste design hay feeders save a lot of money because goats will not eat soiled feed off the ground; in fact, many goats will not eat any food off the ground. This is a wise instinct, since goats are very susceptible to worms. However, it is a costly instinct when they pull big mouthfuls of hay out of the feeder and drop most of it on the floor or ground. Well-designed, no-waste feeders discourage such costly behavior.

The feeder pictured below accommodated 8-10 grown animals. It was free-standing and sat in our outside loafing area. Its rugged construction (painted channel iron and sturdy plywood) gave us years of service. Notice that its lid protected the hay from foul weather. We could put lots of hay in it, but since the goats preferred fresh hay, we made a show of putting in fresh hay at least once daily.

The feeder pictured at the right is a common type of no-waste feeder. Goats pull out a mouthful, and the leaves drop into the trough below. We can recycle the leavings to the does who liked leaves in particular and to kids. These wall or fence-hung feeders work well in small pens and in situations where there is something to

hang them on. Depending on their size, these feeders can hold enough hay for several days.

Another type of no-waste feeder allows the goat to put its head in and eat. Its design usually discourages goats from pulling their heads out while eating, thus cutting down on waste. A typical keyhole-style feeder has 8-inch diameter head openings with a 4-inch wide neck slot. These feeders can be fixed to a wall or free-standing, depending on the need.

Slanting the divider boards on a free-standing feeder forces a goat to tip its head sideways to get out. Our "slant-board" feeder was built for half-grown does. It was 8 feet long, 2 feet wide, and 38 inches high. The slanted boards were spaced 5 inches apart. For adult animals, we built the feeder 3 feet wide with slant boards spaced 6-1/2 inches apart. To use this type of feeder outside, you will need to add a roof on hinges to protect the hay from the elements.

A word of caution: Keyhole or slant board feeders are not suitable for all goats. When a goat puts its head in to eat, it is vulnerable to a more aggressive animal. Use these feeders in situations that you can monitor closely or in pens where you house less aggressive animals.

Below are a few grain feeders you can use. Those that can be hung on a wall or fence are best. Grain feeders placed on the ground or floor allow goats to step in and overturn them.

Hints

"Always allow more feeder space than you have goats. One aggressive goat can block a whole lot of feeder space on its own."

"Old stock tanks whose bottoms have rusted out make great "pens" for newborn kids. Just put them down and fill them with straw."

"We use hay or straw bales as temporary pens for our newborns. We set them together one or two high to form a square or rectangle, and the kids stay warm and draft-free for the first week."

"We find that clean 55-gallon drums, metal or plastic, make the very best grain storage containers."

"To make a hay feeder in a pinch, we cut stock panels to size and attach them at an angle top and bottom to the stock panel fencing with double end snaps. We then put flakes of hay between the feeder panel and the fencing. It's quick to make and not much more wasteful than other types of feeders."

"Old chest-type freezers can make good grain storage."

"Our kid pen fences are made from combination stock panels. To feed our kids grain, we hang our three-foot portable grain feeders on the outside of the fence, and the kids put their heads through to eat. This cuts down on our parasite problem in our kids because they don't constantly have their feet in their grain pans. For our grown does, we place the feeders on their side of the fence (as seen above) if their heads are too large to get through the stock panel openings easily."

Housing and Feeders: Hints

"Provide a sufficient number of feeders so that each goat has its place. Then allow even more space for aggressive goats. Give bossy goats a separate feed pan or hay feeder to avoid fights and subsequent injuries."

"For goat-tight gates, attach the hinges on the gate so that it swings inward, toward the goats. Attach a heavy spring to snap the gate shut after you walk through. Use a latch with a security bar that goats can't open. Then put the latch on the outside, about halfway down, where the goats can't reach it. If you are exceedingly lucky, or your goats are on the dim side, they may not get out!"

"A wooden pallet makes a nice "goat island" out in the goat yard. The Nubian kids pictured here are comfortable and dry, even when the yard is muddy. Goats should not be expected to lie on wet ground or damp bedding. Kids enjoy the "goat islands" and spend more time outside."

"If you're building a new barn, when the trenching is being done, at the same time that you're laying the water line, run a separate trench for electric line. This line should be shielded in plastic pipe. This way, you can put in lights, a stock tank heater, or other amenities requiring electricity later on."

You Can't Keep a Good Goat In, or Can You?

Some General Thoughts on Fencing

First, I would like to go on record as saying that when it comes to fencing in goats, there is no substitute for four feet of woven wire with one strand of barbed wire underneath and two strands on top. However, there are times when other types of fencing will do the job.

For the last two years on our farm, we experimented with all kinds of fencing: plastic, electric, woven wire, and stock panel. Of the four, we found only one to be without merit: the plastic fence. The goats ate it! Not all of it, but enough that the fence eventually became so weak that the goats took an "unescorted tour" of the farm.

The most expensive fence we tried was made of stock panels, but they were also the strongest, doing a good job of keeping the does and bucks apart during breeding season. The does could tease the bucks by rubbing on the fence, trying to back through it, and doing all the obnoxious things does in raging heat do on one side of the fence, while the bucks would paw, stand on, bounce off, and generally try to destroy the fence on the other side. In the end while the stock panels had a few waves and dents, they held in fully grown, aggressive bucks in all their raging lust. (We found that any problems we had keeping goats on their own side of the fence occurred when our gates were not secured properly, not because of stock panel failure.)

We finally did get the whole perimeter of the farm fenced, and we took our own advice, using woven wire and putting up the fence "the right way." We built the fence with 8" diameter treated posts set in concrete every fifty feet with 6'6" steel posts set every ten feet. Corners were constructed with two treated posts on each line with bracing and tension wires between them. We used No. 9 wire, and we attached it to the posts with five wire clips. And we put a strand of barbed wire under and two on top.

All this fencing took a lot of time, sweat, mashed fingers, and money, but we really feel more secure now about our goats staying in and the stray dogs and coyotes staying out.

And Then There's Electric Fencing

We spent a great deal of time and money experimenting with electric fencing, and what follows is a summary of what we found out.

Fencing

First, we found that you do not always get what you pay for. The imported fencers are expensive. We tried a number of these and found that they did not do any better job than the less-expensive American-manufactured fencers. In fact, some were down-right cruel, in as much as they could really burn a goat's hide.

When we tried to find inexpensive fencers we ended up with cheap products that would just barely do the job for the first week and then quit working entirely!

After accumulating a pile of chargers, netting, insulators, posts, testers, and wire, we ended up with two results: the right way to build a sturdy electric fence and a pile of junk that we buried behind the barn.

It would have been nice if we could have just strung one line of wire around the area we wanted to keep our goats in, tacking the wire up onto trees and old bent fence posts as we saw the fellow up the road do to keep his cows in. Unfortunately, goats are just too smart for that (as anyone who has tried this last method will undoubtedly agree).

So after lots of experimenting, here are some guidelines we have come up with for building a good, "almost-escape-proof" electric fence.

First, all corner posts should be strong, either wood posts sunk into the ground three feet or steel T-posts driven two feet in the ground. These posts are the ones that really take all of the burden of the wire when it is being stretched.

The line posts (those that go in a line from one corner to another) can be made from any small diameter (say 1/4" to 3/4") material. We have used re-bar, fiberglass, or regular metal electric fence rods. You can also use T-posts or wood posts with special insulators. These line posts should be placed every fifteen feet and do not need to be sunk in the ground quite as far as the corner posts. Their job really is to hold up the wire. The other important consideration is to keep the fence line straight; otherwise, the wire will try to pull the posts into a straight line.

The next step is to get the proper insulators for your electric fence, and that is not hard. All you need to remember is to use the correct insulator for the type of post that you are using. There are insulators for T-posts, wood posts, and round rods. The corner insulators are a bit different. They can be used with any type of post since they are designed to be tied off from the post with a loop of wire. The electric wire then goes through it.

Now for the wire. There are, basically, three types of wire that you can use: 14 gauge iron wire, poly wire, and poly tape. Taking all things into consideration, we like the poly wire because of its light weight, ease of handling, and lower price. The iron wire is heavy and kinks. The poly tape is very expensive. So for our electric fence, we used poly wire strung at four different heights, 8", 16", 24", and 34" above the ground. Believe me, this is one of the secrets to having a good electric fence. You have to use more than just a couple of strands to keep goats in. And this spacing between the strands makes it hard for goats to stick their heads through and also gives enough physical resistance to hold them back until the electrical "zap" tells them that they are in the wrong place. Again, the real secret to a good electric fence is to string multiple strands of wire tightly.

The last and most expensive component of the fence is the charger. We have used three different types of chargers on our farm, and here are some observations on each one.

Out in our far back pasture, we used a solar charger. The advantage to using it here is obvious. It does not have to be hooked up to any power source; it just sits there and soaks up the sun's rays and then converts them into electrical energy. Of the three chargers we used, this is the only one that does not have a meter on it to indicate whether it is functioning properly. There is just a small red light that pulses on and off. This charger can maintain enough stored power in it to run for 21 days of total darkness! We always checked it at night because we could see the red light flashing from 500 feet away; this saved a long hike out to our back forty.

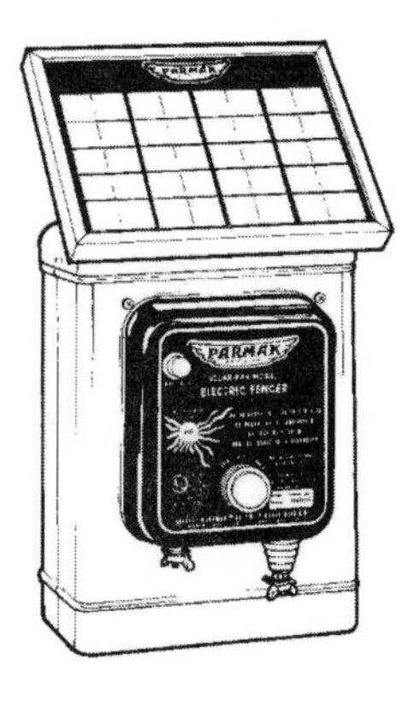

The charger we used to keep the bucks in was located in an implement shed out of the weather. We had plugged it in to a 110 volt outlet. It had a meter on it that told us what the status of the fence was. We could tell whether the fence was charging properly or whether there was a small short or a large one. This charger was the main one we used on our farm. We used it to keep the bucks in their pen, and we also used it on perimeter fencing on the separate pastures for the does. We chose this charger because it was powerful enough to charge fifty miles of fence and did a good job all the way around.

The last charger we used was located in the pasture where we kept the milking does. It charged the fence which surrounded the orchard. This charger was wired to a 12 volt car battery, and it too had a meter that indicated the status of the fence. Like the solar charger, it could be used in situations where there were no electric outlets, yet it was less expensive than the solar charger. You can purchase chargers like these from Caprine Supply or local farm stores. Their UL listing gives us a lot of confidence that they are quality products that have been tested and given a full one-year warranty.

Finally, the last secret to having a good electric fence is to have a good ground. The ground connects the charger to the earth. We have found that an eight foot rod driven into the ground works best. However, if you run into the same problem we did and have only three feet of dirt on top of six feet of rock, you can use two three-foot rods driven into the ground to get the same effect or follow the advice in the "Ground Rods" section which follows .

All in all, it would have been a lot easier to tie our electric fence wire from tree to tree, but we sure got a lot more done not having to chase our does home every day; our orchard thrived, and our does were being bred to the bucks of our choice, not theirs, because we installed our "proper" electric fences.

The following information is reprinted with the permission of the Parker-McCrory Mfg. Co., makers of Parmak fencers. We thank them for allowing us to include this information here.

Fencing Mileage Guide

When it comes to fencing property, one of the hardest problems is deciding just how much fencing you will need to enclose the space you wish to fence. Industry experts use the following chart to help estimate the materials needed. Notice that they use measurements in feet, miles, and rods. We provide the Conversion Table to help you as you estimate your fencing requirements.

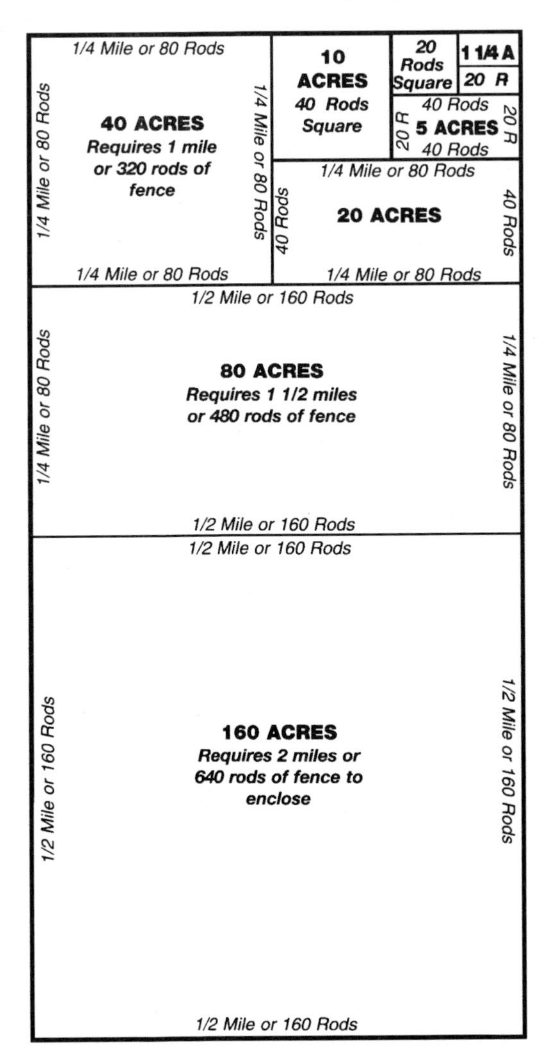

Conversion Table

1 rod	=	16 1/2 feet
10 rods	=	165 feet
20 rods	=	330 feet
1/4 mile	=	1,320 feet
1/4 mile	=	80 rods
1/2 mile	=	2,640 feet
1/2 mile	=	160 rods
1 mile	=	5,280 feet
1 mile	=	320 rods

Ground Rods

For ground rods, use 3/8"- 5/8" diameter eight foot long copper clad rods, and drive them a minimum of six to eight feet deep into permanently moist earth. Sandy, rocky, or clay soil conditions can cause poor grounding. You may need to use additional ground rods when placing fence in these soil conditions. If a fence is not properly grounded, your fencer will be damaged, and there will be decreased shock along the fence line.

If you can not drive a ground rod eight feet deep, drive it in at a 45° angle as deep as you can. Drive four or five ground rods in a circle (like the spokes of a wheel), and connect all ground rods with ground wire and then back to the ground terminal on the fence charger.

Other guidelines on ground rods include the following:

Install the first ground rod within twenty feet of the fence charger.

Do not use painted fence posts or any painted metal rod for a ground rod as paint is an insulator and will not conduct electricity.

Do not use utility ground or water pipe to ground your fence.

Do not install fence ground rods within fifty feet of any utility ground system as this will cause a poor ground condition.

Replace ground rods approximately every two years, as they rust under ground and eventually will not provide a good ground.

This article originally appeared in the Dairy Goat Journal *in 1996, and we reprint it here with our most sincere thanks to Dave Thompson, publisher.*

I Dreamed of a Genie

by Joan Vandergriff

One of my recurrent dreams when I had lots of goats involved a genie. On certain days the genie made goats disappear—some days just individuals, other days whole strings, families, barnsful. My genie also made goats appear—the perfect "100" (the Bo Derek of goatdom!), the perfect "100" that milked 3 gallons a day, the perfect "100" that produced whole litters of other perfect "100s" every year, including the perfect "100" buck who reproduced the perfect "100" offspring no matter whom he was bred to.

Every once in a while, my genie seemed to come through for me, but as I visited certain other breeders' facilities, it seemed that their genie worked a whole lot harder than mine did. When touring their facilities I'd see hundreds of goats relaxing in palatial splendor, while milking took place in parlors that featured state-of-the-art equipment, air-conditioning, and all the amenities of a first-class hotel.

Coming home, I'd think that my genie had been loafing. While my goats looked okay and didn't live on Tobacco Road or even Gilligan's Island, their shelter and milking facility looked nothing like the ones where a genie had fulfilled every dream.

My goats were scattered in various outbuildings, and our milking parlor violated many principles of efficiency: nothing, neither equipment nor tools, was in its place; in fact, most of my goat stuff didn't even have a place. My goats had the run of the farm, coming and going as they pleased not as I wanted them to, and as for achieving efficiency, I couldn't even be a clock-watcher because I didn't even have a clock in the milkroom!

Luckily, my story had a happy ending. My husband and I had bought a piece of property where we planned to build a house and goat facility. We had the chance to design a facility that incorporated many features that made our goat operation highly efficient. While I dreamed of what our genie could do for us, Jeannie seemed to have abandoned us, so we had to do it all ourselves.

For any large project such as this one, we found it pays to put in lots of time up front.

"I Dreamed of a Genie"

Plan, Replan, and Plan Again

Like the old real estate joke, the three most important things to remember when you are thinking about remodeling or building is to plan, replan, and plan again. In our case, to do this, we first went to look at a number of goat operations, and we didn't restrict ourselves to breeders with herds the size of ours. We looked at both the Taj Mahals as well as the Skid Rows of goatdom. You can never tell where inspiration and good ideas will spring from. We made sure that we visited both during milking time and when goats were relaxing out in pastures and barns. We watched how the goats moved about the facility, how the chores were handled, how the goats and chorepeople interacted. We took plenty of notes and even more pictures.

What were we looking for? Any kind of ideas for housing and milking facilities, maintenance, milking and other chores that might just be applicable to our situation. Just because a breeder had more goats or more money, we didn't rule out a scaled-down, less-elaborate, or roughened version of an appealing idea. We got names of suppliers for materials and equipment and asked for any advice the breeder was willing to give us.

Make Lists and Check Them At least Twice

We then made a list of everything our facility had to have. In order to make sure we included everything, we made lists by function (milking parlor, holding pen, kidding pens, kid pens, buck pens, etc.) and by type of animal we intended to keep (for does we needed a milking parlor, holding pen, clipping area; kids needed a newborn area, intermediate pen, and weanling pen, etc.) By making lists in two different ways, we were more likely to remember all the various types of facilities and pieces of the project we needed to include.

Since we were building from scratch, we could skip the next step; however, most people have some structure to work with. If you have an existing structure, you'll need to analyze what you do have, and how it works, how it would need to be altered or rebuilt to fit the list you just developed. Pretty soon you'll be faced with your first very important decision: to get the results you want, should you build from scratch or rebuild or remodel an existing facility? Unless your present facility screams for instant demolition (e.g., the goat police have placed a "Condemned" sign on the door) or you must build from scratch for other reasons, you can't make this decision without making two estimates: what it would cost to build from scratch and what it would cost to remodel the existing structure.

Do You Really Need the Taj Mahal?

While your bank balance will ultimately dictate how much you spend, it's also important to think about how grand a facility you really need in order to do what you want to do. Start with a basic question: will your facility be part of your marketing program? If so, then you're probably going to be building a showcase for goats and perhaps goat products. This may end up costing a whole lot of money. Another consideration: Do you need to pay for your facility out of goat sales and product profits? The answer here may give you your budget's bottom line. You'll need to remember that most businesses amortize capital expenditures over a number of years. In plain English, this means that you can spread the costs over a number of years. Talk with your bookkeeper or accountant if you need more advice on this aspect of the project. But be sure to consider the advice you're given carefully. On paper, you may be able to spread the costs of the project over a number of years, but you'll most probably need to have the cash to pay for it up front! Don't overextend yourself.

Most of us want to build a facility that works efficiently for the least amount of money. Once you know how much you can spend, it's time to decide where you need to put the majority of your budget. If you are building a complete facility (barn, pens, milking parlor), you need to decide if you'll need to concentrate your money in one area over another. If you are remodeling, the answer may be dictated by the existing conditions. You may decide that in order to have the milking parlor you have dreamed about you'll have to forego other niceties in the barn. Early on, decide where you want to concentrate your money. Most of us can't have it all right away.

Even before you decide whether to remodel or build a new structure, take the time to draw up rough plans. You don't need to be an architect or even be able to do more than hold a pencil, pad, and tape measure. However, skip this step and you'll head down a road where even your genie can't help you. If you have an existing facility, begin by making a drawing of what you already have in place. Then put this drawing aside.

Dream On

Next, rough out the basic parts to the facility as you would like them to be. Don't skimp on space, and do put in all the features you would like to have as if your genie were doing all the work and paying for it. By making this "dream plan" before you make your actual plans, you'll be surprised by the number of features you may be able to include that you thought were out of the question.

As you design your spaces, consider the following:

- Having everything under one roof will greatly increase the efficiency of your operation. For example, running does from one building to another two times a day consumes time. Also, in most climates you will need to roof the runs between buildings and build a roofed holding pen outside a separate milking parlor. If animals are all under one roof, you may be able to feed everyone at the same time, increasing efficiency. Once we're in the groove of a task, it's usually more efficient to service all animals than to start, stop, start again, and stop. Momentum seems to increase our speed.
- And while we're on roofs, consider building in skylights. You can easily place transparent panels instead of metal ones at appropriate places, and you'll gain lots more light inside and save on electricity.
- Make your side walls as tall as you can. Remember that winter bedding can cut headroom down drastically, necessitating more frequent cleaning, and in the summer heat, you and your goats will appreciate the free air circulation that high ceilings provide.
- Barn doors are always a problem. Rolling doors or even plain doors get stuck on built-up bedding. Consider a roll-up type garage door when planning your doorways. You can always slide them up easily (bedding height isn't a problem), and on cold or rainy days, you can adjust their open height as you wish. You can't do this with other common door types.
- Dirt floors in the barn will always make your work easier. Dirt floors allow moisture to drain away, make a more comfortable base for goats, and need less bedding. If you own a concrete plant, save it for walls, milkrooms and aisles.
- Don't forget about feed and equipment storage. In general,

keep feed and equipment stored as close to the animals as possible. That means that if you have everything under one roof, you should plan for adequate hay and grain storage. You may not need a full year's supply of hay stored in your goat barn, but you don't want to be hauling hay from a far barn every day or two. If you have animals housed in separate quarters, be sure to have hay, grain, and equipment storage space in each facility. Again, you want to be able to walk into your buck or kid barn and do all your chores without having to haul hay, grain or equipment.

- Remember the "what comes in must come out" rule. Try to include a separate entrance and exit door wherever animals move from place to place.
- Figuring out just how much space you need is tricky. Remember that most breeders outgrow their facility the moment the paint is dry, so think about how large your herd is likely to be in the future and plan accordingly. A rough rule is to provide 25 square feet (a five foot by five foot area) of loafing space for each goat you plan to house. Goats can live in less, but they'll be happier and healthier in more, and your cleaning and other chores will be easier the less crowded your goats are. Remember that kids have special requirements. Usually, you want your newborns separate from older kids. Young kids just can't compete with older ones. We keep our kid pens flexible. We always have separate pens for newborns and weanlings. Kids between two weeks and two months are grouped according to how large and aggressive they are as well as how quickly they feed.
- Design all exits and walkways at least four feet wide. This is the minimum width that cleaning equipment (a garden cart, Bobcat, etc.) can pass through. If you're using a different type of equipment, then build exits and walkways wide enough to accommodate it.
- Design your pens for flexibility. You should be able to segregate animals and turn small pens into larger ones easily. A good trick is to design everything in multiples of sixteen feet. Since a hog panel is sixteen feet long you can use

them to define one large space. Attach each panel to metal fence posts with fence clips. Join panels together at corners with cable clamps. You'll usually not need a fence post at the corners. To split pens, simply run a hog panel across and attach at the ends with cable clamps. Use a fence post in the middle to provide more permanent stability. Temporary doors can be made with half lengths of hog panels. Use cable clamps for the "hinges" and an appropriate tie top and bottom to open and close the opposite end. The beauty of this system is that you can always split or open pens as you need to. It's easy and merely takes (un)fastening two cable clamps at each end.

- Anywhere you are likely to place water buckets, build a drainage sump. Do this by digging out a large hole and filling it with rock. Even though it takes work, you'll be pleased by how well it keeps water drained away. To keep bedding from clogging the sump, you can build up lips around the sump with wood and as part of daily chores, remove any stray hay or straw.
- For maximum efficiency, consider investing in some type of automatic watering system. Anyone who has spent time doing chores knows how heavy those buckets are to fill and carry. Plus we all have animals that love to push them over, dumping water, and rolling buckets down the hill.

There are a number of ways to go with automatic waterers. The most common are the "pig waterers" or nipple valve waterers that attach to your faucet. To drink, the goat places its mouth on the valve and "tongues it." I'm not wild about these only because goats need time to learn how to use them, they have to be persistent to get a sufficient amount of water, and there's always one goat that stands there with its tongue against the valve, allowing it to spew water on the ground, making a mudhole. If you use these waterers, be sure to build a drainage sump under each one.

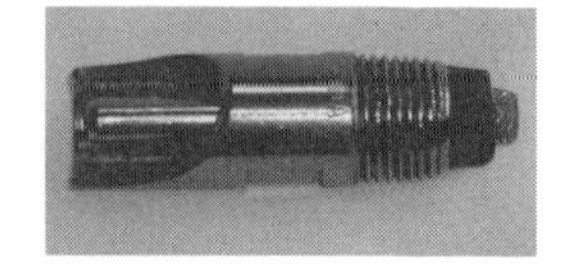

Small bowl-type waterers work well, though they don't have much capacity. The best we've found are horse waterers. They're expensive, but they have a large capacity, rarely leak, and have a heating element that works great to keep water from freezing in cold climates. If you can't afford these throughout your facility, think about putting in a couple where you'll get the most benefit. You won't be sorry.

- Wherever you may need to use electricity, put in an electrical outlet. If you use the horse waterers above, you'll need an outlet for each one. Be sure to provide electrical service near your kid pens so you can disbud kids right out of the pen. Decide where you'll be doing your clipping and put outlets there also. Remember, it's much easier and cheaper to install electrical outlets when you're doing your initial building or remodeling rather than adding them later.

In addition to the above considerations, you'll probably have lots more of your own. Once you have your "dream plan," price out the costs of building it from scratch and by remodeling (if this is possible). You may find that remodeling is a more expensive route. By drawing plans and estimating both ways, you'll find out which way makes the most sense for you.

Back to Reality

Like most "dream plans," we usually can't afford them (unless you have a devoted genie), so the next step is to make some hard decisions about what are absolute necessities and what are niceties that you could, if you had to, give up. You may even need to make some decisions (which you'll need to stick to) about reducing the size or growth of your herd. Keep working at your plans until they and your budget match up. Remember that most construction projects go over budget for any number of reasons, and there's no reason to expect that yours will come in on target. Count on your project costing more than your plans call for, and budget accordingly.

As you shop for materials, get lots of bids and don't get stuck on any one item or brand. Be flexible, look for lower cost options that will work just as well, and avoid special order items. If you can find a single vendor who has most of what you'll need (even if some items are a little more expensive) talk with him or her about getting everything there. You'll probably be able to get a good price and order items at a good price that may not normally be in stock . Also, remember that your time is worth something in this whole process, so if you have to go to twenty places for materials that are only a little less costly than going to one store that has everything, you're not saving anything. Finally, be sure to ask about delivery costs, especially if you live far away. A good price may not be very competitive once the delivery is tacked on.

Once you have your revised plans and your materials, you're ready to begin. As you go about your project, be sure to keep your final plans in front of you. Remember that any changes you make may affect other parts of your plans, calling for replanning and often increased costs. A good rule is: make no changes once you've begun construction. If you do, be prepared for increased construction time and cost.

Our new facility, when done, was neither the Taj Mahal nor Poverty Flats. Since we did it without the help of a genie (I guess we just weren't "genie-type" people), what we ended up with was a compromise between what we could afford and all the things we dreamed about in a goat facility.

Most importantly, what we built worked well for the goats we had when the project was finished and the increased number of goats we had five years later. It was easy to clean, airy, and bright. We could move goats from one pen to another easily, build isolation pens in five minutes, and break down pens for cleaning in a flash. Chores were efficient, and since everything we needed was located near its use, virtually anyone could walk in and learn to do chores almost immediately.

So when you're ready to undertake your next big building project, you can dream of all the things your genie can provide you with, but in the end you'll get the best results if you plan, replan, and plan again.

Raising Kids . . . An Overview

One of the first decisions you will have to make when your first kids are born is how to raise them. Kids can nurse their mothers or dams, or you can hand raise them, feeding milk in bottles, pans, or a Caprine feeder.

Nursing is easy, but there are disadvantages. First, you may pass on health problems from dam to kids. You must also be sure the dam is producing enough milk to feed her kid or kids, and that no kid is being rejected. Kids that nurse their dams may be a bit wild, so you will need to spend extra time handling them to tame them down.

Some people allow their kids to eat all day long, feeding cold milk free-choice. Breeders who have tried this like it, reporting that kids do very well, and the labor of raising kids is kept to a minimum. The drawback is that you cannot monitor an individual kid's intake, and you must be sure that the milk does not spoil. You may have better luck feeding cold milk free-choice in cooler climates or the winter months than where or when it is warmer.

Many breeders prefer hand raising their kids. If you do this, you will know that each kid gets enough milk; you can control the milk intake, raise kids separately from adult goats, and wean certain kids early if you want to.

If you have just a few kids, you can bottle feed them from birth to weaning. With larger numbers, you will probably start kids on a bottle then move on to some kind of gang feeder where a number of kids can drink milk at once.

At our farm, we follow a rather standard kid raising routine:

Our kids are fed three or four times a day the first couple of days, and then we feed milk just twice a day. They get as much as they will drink, up to 1 1/2 pints a feeding. We give them good leafy legume hay, and a little bit of molasses mix grain (the same grain our does are fed) is provided in feeders by a week of age. We throw out any leftover feed, because kids tend to contaminate it, which can lead to health problems, and vermin gravitate to these leavings also.

When kids are a month old, they really start consuming lots of grain, so we limit grain to a half pound per kid, fed twice a day. After weaning, they get a pound of grain a day and all the hay they want. Weaned kids get mostly grass hay, plus the stemmy leftover legume hay from the does' feeders.

Some breeders feed as much as a gallon of milk a day to each kid if the kid will consume it, and they may not wean kids until four months of age. Other breeders feed a quart of milk a day maximum and wean as soon as the kids are eating grain well, as early as six weeks. Both systems can do a good job of raising kids to be good producing does.

If you have surplus milk, it is cheaper to feed your leftover milk than purchased grain, so heavy milk feeding may make sense. Kids may be raised on cow milk and also on milk replacers, but be sure you use **kid** milk replacer since it is based on real milk solids, not vegetable proteins such as soybean formula.

Raising Kids: How We Do It

At Caprine Supply we get lots of phone calls about raising kids. To answer some of these questions, we have written the following guide on how we raise kids. What follows is not the only way to do it. In fact, you will probably want to do things a bit differently if your goats are CAE-positive, for example. However, after thirty-five years of experience, this is the system we have found that works the best for us and our goats.

First Comes Mom: Effective kid raising begins long before the kid ever hits the ground. The dam must be in very good condition. Because any pregnancy is hard on a goat (remember that she is probably still giving milk through her first three months of pregnancy), make sure that your does are getting quality hay and grain, that their rations are balanced correctly, and that they have received their immunization shots. Observe all of your does carefully. When a doe is pregnant, time can be a critical factor if she has problems. The earlier you catch a problem, the better it is for both her and her kids.

No Kidding Pens: We keep our pregnant does in their regular pens, even when they give birth. We do not put does in kidding pens; we keep these individual pens for the occasional doe who has problems. Does often get nervous when they are not in their regular environment, and does get nervous enough during kidding so they do not need added stress.

People tend to use kidding pens because they are supposed to be cleaner than their regular pens, providing a good environment for newborn kids. Our regular pens are cleaned every week and are clean, so we do not see a problem there. Also, if you keep a doe who is kidding in her regular pen, there is usually an older doe around to help clean up kids if the mother gets involved having another kid or just does not want to bother with her newborns.

Watch Those Goats: We check breeding dates and watch our does closely, so we know the "window" when a doe will kid, and we tape her teats with teat tape a few days before. (For health reasons we do not want any kid to nurse its dam.) From then on, we keep a very close eye on the doe, and we try to be around at kidding time so we can help if there is a problem. We sometimes joke that we can look in a doe's eyes and see that she is going to kid that day. After a few years' experience both with goats in general and your own goats, you will probably develop a feeling for who is going to kid when.

Those First Few Hours: When the kid is born, we either let the doe clean it, if she wants to, or we clean it up and dry its coat, using clean old towels and even a hair dryer in the winter. We then dip the kid's navel in strong iodine. We use strong iodine because it cauterizes the navel cord quickly, lessening the chance that bacteria will travel up the cord. Remember, a wet navel cord acts like a wick, drawing germs up into the kid.

We Feed Pasteurized Milk: We raise all of our kids on pasteurized milk. We do this primarily because we want our kids to be CAE-free. For more than ten years we have listened to opinions from people whose opinions we trust, and we keep coming back to this approach. It works best for us. Our does no longer have any swollen joints, they do not have hard udders on freshening, and a doe raised on pasteurized milk generally milks more than her dam (if the dam was not raised on pasteurized milk). Of course, we are not veterinarians, so we can not say for sure why raising kids on pasteurized milk works so well; however, generally, kids raised on pasteurized milk are healthier and do better than kids we used to raise on their dams or on unpasteurized milk.

If you raise kids by hand on pasteurized milk, there are additional benefits. For example, kids left on their dams drink far more milk than they need, often putting on excess weight and fat. This fat collects on shoulders and in the udder, among other places. And most of this fat never leaves the goat. Like some humans, a fat kid often means a fat adult who rarely milks up to her potential.

In addition, kids raised by hand, bond to you as their mother. This means that they are extremely friendly, they are easy to work with, and they know their place (we are the boss, they are the kid!).

Those First Feedings: If our new kids are healthy, alert and active, we use our hand-held Caprine feeder immediately. The first few feedings are messy: milk all over both the feeder (us) and the "feedee" (the kid), the floor, etc. However, most kids catch on quickly. We let them drink until they have had 6-8 ounces or their stomachs are tight.

The Problem Child: We follow that routine for "normal" kids. What about the problem ones? First, there are "stupid kids." If they are up, alert, and active, but think the Caprine nipple is "poison," we try feeding with the Pritchard nipple on a pop bottle. If it works, great. If not, we put the kid back in its nice warm pen for a few hours. Then we try again. It is a miracle how a once "stupid kid" gets the idea when his stomach is growling! It sometimes takes three to four hours before a kid gets smart. That is okay as long as the kid is in a nice warm spot and does not get chilled.

Then there are weak kids. If we see a kid is not responding as we like it to (it is weak or lethargic), we first rub it all over to stimulate it and get some more warmth into its body. The real trick here is to get the weak kid to nurse. If it does not nurse (and we do not let weak kids wait like healthy ones if they do not nurse), we get out the tube feeder and "tube" them. The more weak kids you handle, the easier it is to decide whether to or when to tube feed kids. We save lots of kids this way, and we do not hesitate to tube those that need it.

Bring 'Em Back Alive: We have had luck with some kids we have found lying in bedding with no sign of life. We take them and immerse them (keep their head out!) in a bucket or sink of warmer-than-body-temperature water. This gets their body temperature up. We then wrap them up like mummies to dry them and tube feed them. In some cases we have used a mixture of 2/3 colostrum and 1/3 black coffee. The caffeine gives them a "jolt."

Out to the Barn: With our normal kids, once they have had their first drink and are dry, we put them out in the barn in our newborn kid pen. We keep a number of kid pens through kidding season, "graduating" kids from one to another as the pens get crowded or the kids outgrow the younger kids being added to the pen.

When It Is -10°F: If the weather is especially cold, we put a couple of sky kennels in the pen for kids to get into. We take the larger kennels apart and get two houses from each kennel. But a word of warning: kids can pile in these kennels so tightly that the ones at the back can suffocate. Be sure to give kids enough room, and do not use the kennels if there is a big difference in the size of kids or there are weak kids in with active ones. You can use bales of hay or straw to build windbreaks in your kid pens. What you want is a warm, dry, draft-free environment.

Graduation Day: We feed our kids on the hand-held feeder for about three days. Then, we put them on the regular Caprine feeder. At first it is a big job making sure that each kid stays on and gets its fill. However, the time you take helping them to master the process will save you incredible time over the spring. Once our kids get good at it, we can feed 40 kids in 15 minutes, including clean up!

Out into the World: When kids are about two weeks old, we move them out of the big barn into our outside kid pens. These pens have sheds, approximately 10′ long by 6′ deep, with a 2′ wide door in front. Each shed has a large yard, 30′ by 50′. The yards have "toys" including cable spools and teeter-totters.

We use hog panels to enclose the yards and hang hay feeders on the inside and grain feeders on the outside of the fencing. The kids stick their heads through the panel to eat the grain.

It is most important that the pens are kept clean. This means changing bedding every four to seven days. Kids that drink milk cycle a huge amount of moisture into the bedding, and dirty, wet bedding means sick kids.

We begin putting out hay and grain almost immediately. We know very young kids do not eat it, but they begin to get serious about week two. We do not feed much grain to our kids mainly because we do not like fat kids. We figure one cup a feeding as the kid grows and is bred. We continue this regime until their last month of pregnancy when we up the ration just a bit. We do feed a lot of good hay. We like to get their rumen functioning quickly.

Weaning Day: We like to wean our kids at eight weeks of age. Every once in a while we will have a kid that needs milk a bit longer to encourage growth. However, a kid that is growing well, eating both grain and hay, should be able to move on to a totally "solid food" diet at eight weeks. Other breeders may feed kids longer (up to four or even six months), but for the breeds we have raised, we find that kids fed lots of milk tend to be "fatties," and we like our kids growthy but trim. The key is to monitor how your kids are growing and make your decisions based on your herd and your individual kids.

Raising kids takes patience and perseverance. But our favorite part of raising goats is watching our kids grow and thrive, turning into milkers that live long productive lives.

Raising Kids on Pasteurized Milk

Why Pasteurize?

There are a number of diseases that can be passed on to kids when they drink raw goat milk. A common one is Caprine Arthritis Encephalitis (CAE), which was identified by research workers at Washington State University. This virus causes arthritis in goats of all ages. It occasionally causes paralysis in kids. In the past, before CAE prevention programs, more than 80% of goats tested for CAE in the United States tested positive, although many did not show symptoms. Check with your veterinarian for more information about CAE testing.

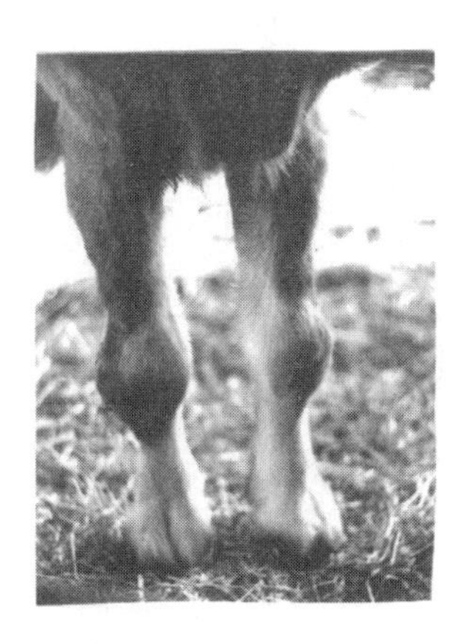

This doe's big knees are probably the result of CAE infection

Diseases Besides CAE

CAE, however, is not the only serious disease passed through raw milk. Mycoplasma and chlamydia have been commonly identified in goat herds. Besides arthritis, they can cause pneumonia, mastitis, abortion, eye problems, and other illnesses. And while these diseases may be passed from infected animals to healthy ones through routine contact, you can improve the health of your kids (and eventually of your entire herd) by preventing your newborn kids from drinking raw milk or colostrum from infected does.

Alternatives to Feeding Raw Goat Milk

To prevent passing on these diseases, you can feed pasteurized cow milk or kid milk replacer, or you can do what we do: raise kids on pasteurized goat milk and heat-treated colostrum.

To feed newborns:

First, newborn kids must not nurse their dams at all. A single feeding of colostrum from an infected doe might be enough to spread disease to the kid. So try to be present when kids are born so you can remove them from their dams immediately.

Of course, it is not always possible to be there. So if a doe seems close to kidding, you can put tape around and over the ends of her teats to keep kids from nursing. Teat tape works quite well. Newborns should be fed heat-treated colostrum or any of the commercial colostrum powders or gels.

To heat-treat colostrum:

You can not pasteurize colostrum because it will turn to custard. The best you can do is heat-treat it. Here is how to do it:

Put colostrum (the larger the batch, the easier it is to control the temperature) in a double-boiler or water-bath, and heat it to 135°F. You must then hold it at this temperature for one hour.

An easy way to do this is to use a thermos bottle. Fill the thermos with 135°F water to preheat it. Warm the colostrum to 135°F. Next, empty the thermos, put in the colostrum, put on the lid, and set the thermos aside for an hour.

Once your kids have had colostrum (we feed ours colostrum for their first 24 hours of life), you can switch them to pasteurized milk.

To pasteurize milk:

Pasteurizing milk is a lot easier than heat-treating colostrum because there is no holding time. Just heat the milk to 165°F, stir it to make sure all the milk in the kettle is 165°F, and you are done.

You can pasteurize milk on the kitchen stove. Fill a tall tote pail with milk. Set the pail on a rack in a canning kettle with some water, and turn on the heat. Use a large dial thermometer that clips on the inside of the tote pail. Some people pasteurize milk successfully without using a water-bath. They say the trick is to make sure you do not use milk from "just-fresh" does. Because it probably contains some colostrum, this milk tends to scorch easily.

Using a home pasteurizer is most convenient. A buzzer sounds when the pasteurization process is completed. Not only can you do something else while the milk is being pasteurized, but you can use the pasteurizer in the barn since it runs on ordinary 110-120 volt current. The Weck-style canner/pasteurizers work exceptionally well for pasteurizing milk.

While pasteurizing milk seems like a lot of trouble, we believe in it. Many breeders have eliminated "big knees" and swollen joints in their herds and have seen fewer cases of pneumonia. When does raised on pasteurized milk freshen, they tend not to have udder edema (hard udder), and their milk production is often superior to that of "nonpasteurized does." Best of all, kids are more robust and healthy, and they grow up to be good milk-producing, long-lived, healthy does.

And in the Future?

Most breeders who sell stock, test their herd for CAE. By knowing which animals are positive for CAE, they can make decisions about culling and how to raise their kids. Once their herd is completely negative (usually accomplished through culling, segregation, and raising kids on pasteurized milk over a number of years), they may go back to raising kids on their dams or with unpasteurized milk. However, any time they bring a new untested animal or CAE-positive animal into the herd, they run the risk of introducing CAE back into the herd. Therefore, breeders of CAE-negative herds usually do not buy outside kids (or purchase only CAE-negative kids) and they practice segregation of any animals they bring in for boarding or breeding.

Using Caprine Feeders

The Caprine feeder Takes the Work Out of Feeding Kids

The "Caprine" style milk feeder really saves time when feeding kids.

The feeder consists of a bucket with kid-level holes, each fitted with a special Caprine nipple and a tube that reaches into the milk. Kids "drink through a straw."

You can feed ten kids at once, and if you have more than ten kids, use two feeders. Using a Caprine feeder, you can feed fifty to sixty kids in less than thirty minutes.

Training kids is easy. Most kids will simply latch on to the nipple and suck, drawing milk up into the tube. If you have a slow learner, fill a quart jar with milk to the level of the nipple so that just one suck will get milk into the kid's mouth. Train one or two kids at a time. Some learn quickly; others are slower, but it is worth the trouble to teach them.

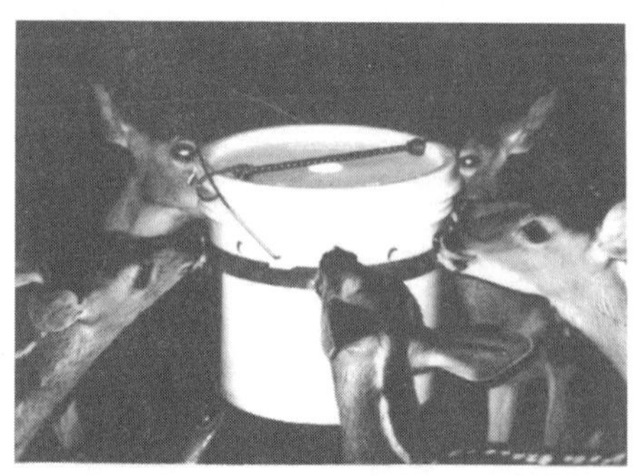

Orphan fawns feeding at a Caprine feeder at a wildlife rehab center.

To give kids a measured feeding, put five one-quart jars inside the bucket. Run a tube into each jar, and feed five kids at a time. Each kid drinks from his own jar, and it is still faster *than* bottle feeding.

To clean the feeder, rinse it thoroughly. Then fill it with soapy water and allow the water to run up the tubes and out the nipples. Finally, run clean water into the bucket and up through the tubes and nipples. Clean the tubes with a tube brush as needed. You can sanitize your Caprine feeder by rinsing it (tubes and nipples included) in a solution of water with a little bleach. However, you will notice that the nipples and tubes will not last as long using a sanitizing rinse.

Caprine Nipple Training Feeder

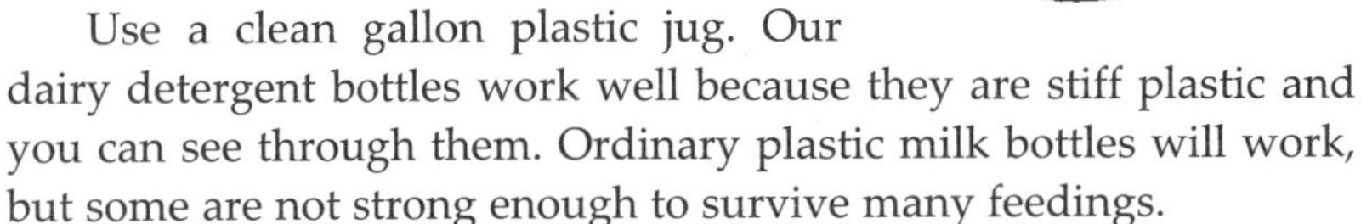

To train new kids to a Caprine feeder, we start them off with colostrum in a pop bottle and a Caprine nipple. Once they are on pasteurized milk, we move them on to a Caprine nipple training feeder.

Here is how to make a five-kid training feeder and train kids to use it.

Use a clean gallon plastic jug. Our dairy detergent bottles work well because they are stiff plastic and you can see through them. Ordinary plastic milk bottles will work, but some are not strong enough to survive many feedings.

To make the feeder, measure about 5″ down from the top of the bottle, and cut about 3/4's around front and sides, leaving the handle intact for carrying. Make five 5/8″ holes just below the cut. Pull Caprine nipples through the holes and snap them in place; attach Caprine tubes to the Caprine nipples.

Feed kids colostrum individually for the first 24 hours.

To train kids to use this feeder, make sure they are hungry for next feeding. Fill the jug with milk to cover the tubes, causing milk to drip from nipples. Start kids one at a time, holding each baby on your lap the same as you bottle feed. Once kids get the first taste of milk, they will suck until they are full. In a day or two, kids will be running to the feeder for their milk.

Hints

"If you must feed kids with pop bottles, a bottle rack system cuts down on feeding time. Bottles sit on a slanted board with the bottle's nipple protruding through a hole in the front board. Note that the bottles lay at an angle, allowing kids to drink to the 'very last drop.' Make sure that the hole in the nipple is large enough for the kid to drink but small enough so that it will not allow milk to drip freely when you place the bottles in the rack. Also, make the hole in the rack that the nipple slides through only large enough for about half of the nipple to slide through. If you do this, kids are less likely to pull the nipple off the bottle. You can either nail the rack to the wall or keep it portable, depending on your needs."

Raising Kids: Hints

"I can't count the times I've come into the barn to find a badly chilled newborn kid. Warming it quickly is the key to saving these kids. One way I've found that really works well is to put the kid in a pail of warm water, then dry it first with a warm towel, and finally use a hair dryer to dry its coat thoroughly. When the kid is all toasty, I put a goat coat on it. If the baby is only a little chilled when I find it, I put it inside my sweatshirt and carry it around close to my body (as if it were in a papoose) until it's warmed up."

"We put a baby sweater on kid goats to keep them warm in the winter if it's especially cold. Just button the sweater over their backs.

"After we dry the ears of newborn Nubians, we dust the ears with baby powder to prevent their sticking together."

"Instead of automatically tube-feeding weak kids, try the Pritchard flutter valve nipple first. Tube-feeding can be hard on kids, and the way the Pritchard teat is constructed, even very weak kids may be able to suck enough milk to get them off to a good start."

"We never have a problem switching our new babies from a bottle to the Caprine feeder because the only time they ever get a bottle is for their colostrum feedings. After that they go right on a hand-held Caprine-style feeder. It's fast, and the kids love it, even 12-hour old babies."

"If you start babies out on a pop bottle nipple and then switch them to the Caprine nipple, some babies reject the new nipple. We've overcome this problem by feeding them colostrum from a bottle topped with a Caprine nipple rather than a pop bottle nipple. That way, the only nipple they've ever had is a Caprine nipple, and they're used to its feel when we switch them to the Caprine feeder."

"The Albers-style nipple fits on some one-quart juice jars. I like to use this rather bulky jar for feeding older kids because it's easier to clean than pop bottles."

"To identify newborn kids, use i.d. neck bands with the name of the mother, date of birth, father's name, and any other information you need. Use a pen with permanent ink and wrap the surface you've written on with clear tape for longer visibility."

"To identify newborn kids, I keep animal crayons in the barn. Once a kid is born and dry, I immediately mark it and its mother with the same color crayon. I re-mark them as needed until I get around to tattooing the kid. With color breeds such as Saanens or Toggenburgs, I don't depend on my instincts to pick out a certain mom's kid. I mark them immediately so I'll be sure who is who from the beginning. I once had so many kids born at the same time that I ran out of different crayon colors and had to mark one mom and her kids orange on the front leg and another orange along her flank."

"For times when you don't have colostrum on hand, keep some heat-treated colostrum in your freezer. You can fill ice cube trays or put the freshly heat-treated colostrum in either seal-a-meal bags or Ziploc freezer bags. Then you can thaw colostrum in its own pouch in warm water (don't use hot water or a microwave as either can destroy the antibodies) whenever you need it. "

"We take our large size shipping crates and put one half in each baby pen. Kids pile in there to sleep. Just don't put too many kids in one pen because the kid on the bottom can get smothered by the weight of the others."

"I like putting teat tape on my does after they freshen because between milkings, I can put the kids out in the pasture with their moms. This teaches kids better foraging and eliminates extra feeding in the barn."

Disbudding: An Overview

Most of us have seen pictures of mountain goats butting each other, and such butting behavior is common in domestic goat herds also. This is no problem if none of your goats have horns. However, horns are dangerous, and a horned goat can injure other goats and humans as well. Children are at special risk around horned goats as their faces and eyes, in particular, are at just the right height to be vulnerable to the sharp ends of a goat's horns.

A Toggenburg buck with a full set of horns. While he is a handsome fellow, he still should have been disbudded as a kid.

There are other reasons not to have goats with horns: it is often difficult for a horned goat to put its head in a keyhole or other no-waste type feeder. And even if they manage to get their heads in this type of feeder, they often are stuck there until some human comes along to free them. The same is true for horned goats and fencing. For some reason, a horned goat always finds the grass greener on the other side of the fence and never figures out how to extricate itself once its head is through the fence. Also, a goat will learn to use its horns like a crowbar, pulling boards loose, tearing apart wire fencing, and breaking open feed bags.

Safety is always the wisest course. It takes very little time to disbud a kid when it is young. To remove horns from older animals is a bit more difficult. Most removals should be left to your veterinarian. However, some owners of horned goats have successfully reduced the danger and hassles by sawing down their goats' horns, keeping them short and blunt. This lessens the danger but does not eliminate it.

Disbudding: An Overview

Most goat breeders remove potential horns when goats are babies. They do this using a hot disbudding iron or, less commonly, caustic paste.

While on first glance disbudding kids may seem a cruel practice, if you think about the environment where we keep most domesticated goats today, you will soon see the sense in removing horns from kids and not keeping adult goats with horns. Breeders of dairy goats have historically disbudded their kids soon after birth. Breeders of other types of goats are following this practice in greater numbers than in years past.

Naturally Hornless Goats

Some goats are naturally hornless or polled. They have small hair-covered knobs on the head instead of horns. If both parents originally had horns when they were born, their kids will always be horned. If one or both parents are hornless, the kids can be either horned or hornless.

Naturally hornless goats should not be mated together because they may produce kids that are hermaphrodites (bi-sexed) or less fertile than goats born with horns.

Genetically, horns are recessive, and the polled gene is dominant. However, most breeders have tried to eliminate polled animals from their herds because of the fertility problems, so you will see fewer polled goats than the rules of genetics would lead you to believe exist.

At birth, it might be difficult to tell if a kid is hornless. (Remember: if either parent has horns, its offspring will also be horned, so you will only be looking for hornless kids if one of the parents is hornless or polled.) Look at the hair that lies over the area where you would expect to find horn buds. On hornless kids, the hair there will be straight (rather than swirled for horned kids), and the skin over the horn bud area can be moved back and forth. You will see some horn growth on Swiss breed kids within a week. On Nubians, you might have to wait as long as a month. Eventually many polled animals develop small knobs in place of where they would normally grow horns.

How to Disbud Kids

The most effective way to keep horns off goats is to disbud kid goats with a hot iron before they are a month old. Usually you should disbud kids at four to ten days of age.

A proper disbudding tool should have a tip 3/4″ to 1″ in diameter.

You can use a piece of pipe heated with a blowtorch to burn the horn buds, but the electric disbudding irons are a lot more convenient.

Using a disbudding box to hold your kids during disbudding helps keep kids immobilized and allows you to put your attention where it needs to be, applying the hot iron to the proper place on the kid's head.

We prefer to disbud our kids when they are a few days old and the horn bud has grown enough that it can be clearly felt. At four to ten days of age, the kid is strong, the skull has thickened, and you are less likely to accidentally disbud a naturally hornless kid.

Buck horns grow faster than the horns on doelings. Nubian horns grow more slowly than the horns on other breeds. A Swiss breed kid's or LaMancha buck's horns may be visible at four days of age, while the horns on a Nubian doeling may not emerge until she is a month old.

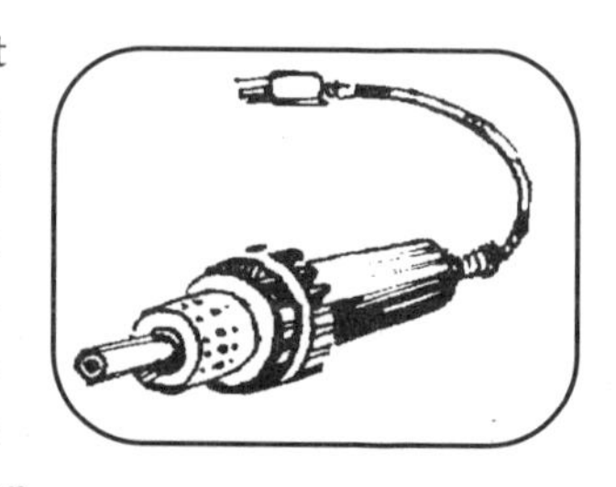

How long do you burn with the iron? It depends on how hot the iron is and how large the horn bud is. With a lower-watt iron (the Lenk 125, for example) we usually burn each horn bud 15 to 20 seconds. With a hotter, higher-watt iron (Lenk 200, Rhinehart X30 and X50, the Dual, or the Goat Dehorner) we usually burn 6 to 15 seconds. We test the iron on a scrap of soft wood to see how quickly it burns a black circle. If the horn bud is large, we may burn once, let the head cool down, and then burn again.

A proper burn leaves a copper-colored ring over the horn bud. It is best to clip around the horn area before disbudding. This enables you to watch the color change and eliminates the smell of burning hair.

Be sure that your kids have been vaccinated for tetanus before you do any disbudding or scur removal.

Scurs are malformed bits of horn which grow when all the horn root has not been destroyed. To help discourage scur growth, study the horn bud before you start. Be sure to burn to the outside edges. Move the handle of the iron in a circle while you burn so the tip covers all of the horn bud.

If the kid has visible horns, you may need to remove the horn tip first. Burn around the base, and the horn will come loose like a little leather cap. Then burn the root. Sometimes we slice a small horn off with a wire saw; we then burn the horn area quickly to stop any bleeding.

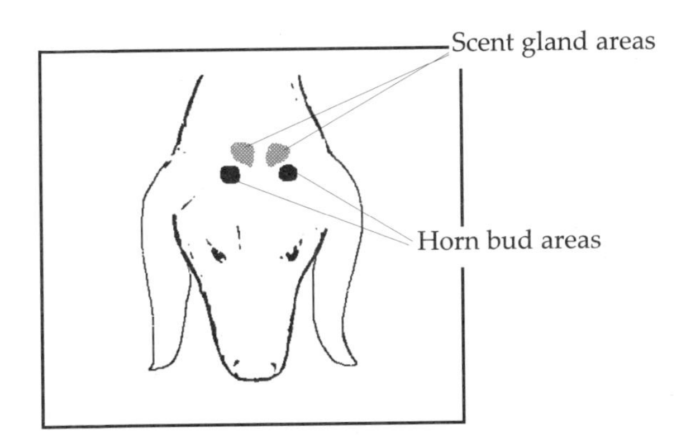

To descent buck kids, burn the scent glands when you disbud. These glands are located behind the horn buds and slightly toward the center. A descented buck is never completely odor-free, but he will be much less obnoxious smelling in breeding season.

Check disbudded kids every two to three months for scur growth. If you find even a very tiny bit of horn, heat up the iron again and burn the horn area. That small scur may turn into a large problem by the time the goat is an adult. It is easy to eliminate scurs when the goat is still young.

How to Disbud Kids

Some goatkeepers use disbudding pastes or caustic sticks to kill horn bud growth. These substances are potentially dangerous because they can run off into a goat's eyes or get rubbed off on another goat. They must be used with care. People who use caustics successfully restrain the kid for a half hour or longer after applying it. Some hold the goat in their arms while they watch the evening news on TV; others lock the kid in their kid-holding box.

Differences in Disbudding Irons

All the disbudding irons in the Caprine Supply catalog are intended for use on kid goats; however, several of them can be used to disbud calves.

The more expensive irons, rated at more watts, get really hot. They are almost "red" hot if viewed in a dark room. They burn the horn bud very quickly, and they hold the heat very well, so you can disbud a number of kids, one after the other.

With a very hot, heavy-duty iron such as the Lenk 200, the Rhinehart X50 or X30, the Goat Dehorner, or the Dual, you would burn each horn bud 6 to 15 seconds. You can also disbud several kids in a row without waiting in between to let the iron reheat.

The less expensive disbudding iron is rated at lower watts. It does not get as hot, nor does it hold the heat as well as the more expensive irons. It will get hot enough to disbud kids if you use it properly. When using the lower wattage iron, you must burn each horn bud longer and allow at least one minute between burnings for the iron to reheat.

Both the Rhinehart X50 and the Dual can be used to disbud calves. Just remove the special Caprine tip from the X50 and it becomes a calf disbudder. The Dual has two sides; one side is specifically for calves.

The more expensive irons are more reliable. Most new irons are guaranteed for a year; however, the manufacturer must repair or replace them for you. This is time consuming and annoying when you need to use your iron.

Some breeders report they do a better job and have fewer problems with scurs when they use a heavy-duty iron. However, several breeders have used the lower-cost Lenk 125 iron for five years or more with good results and no repair problems.

The secret for good disbudding is burning long enough to destroy the horn root and moving the iron around to get to the edges of the horn bud. You must also allow the iron to reheat between use.

Non-Electric Irons and Disbudding Boxes

Non-Electric Irons

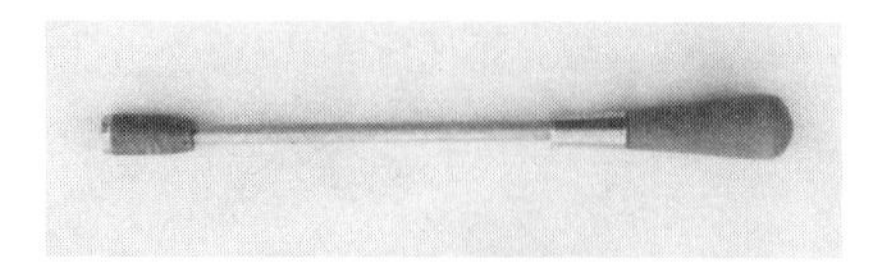

The non-electric iron is an efficient, inexpensive way to disbud one kid or a lot of kids. You will need a blow-torch to heat it with. Hold the tip of the iron in the blowtorch flame until the tip barely glows. Burn each horn bud about ten seconds. Reheat the tip in between each burn. The blowtorch can be left on while you work, allowing you to disbud one kid after another. The tip on the Caprine Supply non-electric iron is the same as the one on the Rhinehart X50. The wood handle makes the iron lightweight, yet sturdy. And there are no electrical parts to burn out.

Building a Disbudding Box

Good dimensions for a kid holding or disbudding box are 5" wide, 24" deep, and 16" high. If you disbud many oversized kids, you may prefer an 18" high box.

Use 1" lumber for the top, bottom, and ends. Use 1/4" plywood or masonite for the sides. A 2" x 4" block inside helps support the kid. The lid must be a half inch wider than the bottom and end boards. The head opening is U-shaped, 3 1/4" wide at the top and 4" deep.

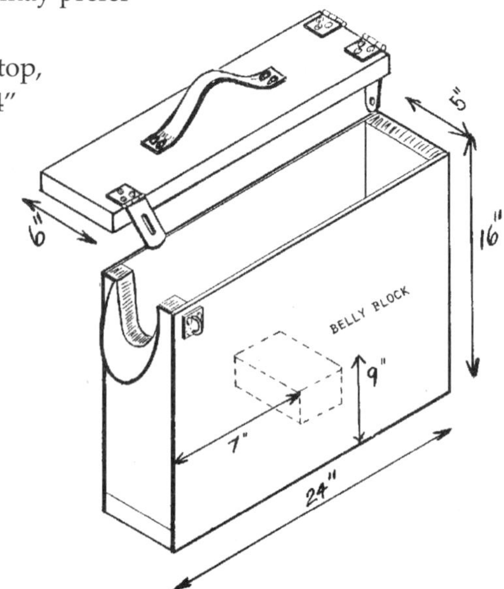

Hints

"When I use my kid holding box for disbudding kids, I put a towel where the kid's neck hits the aluminum headpiece. It cushions the kid's neck a little and makes me feel better about the whole procedure."

"After disbudding a kid, I put a handful of crushed ice on its head where I burned it. It cools down the head quickly, and the kid really appreciates this. The sensation of cold also gives them something to think about besides their burned horn buds!"

"Before you use your disbudding iron and in between each kid you disbud, brush the surface of the disbudding iron's tip with a wire brush. This removes built-up carbon particles and allows the iron to get hotter and hold the heat better."

"Here's a trick for heating the non-electric disbudding iron initially or keeping it hot between disbudding kids. Prop a can (like a coffee can) off the ground with two bricks. To heat the iron, put the iron in the can, and position the heating torch so that the flame hits the disbudding iron as it rests in the can. This will help the iron heat faster."

Toggenburg buck with an unsightly scur. Note that the buck can no longer lift his right ear because of this scur.

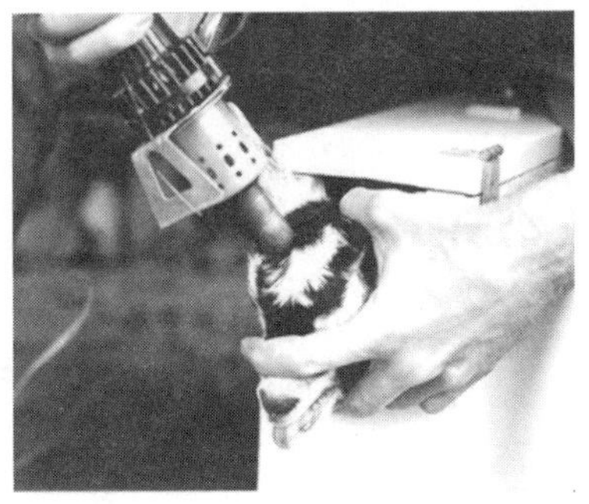

Alpine kid being disbudded with a Rhinehart X30 iron in a disbudding box.

Disbudding

Tattooing: An Overview

To permanently identify dairy goats (and most other goats, too), we use tattoos. These tattoos are usually placed in the goat's ears. Since LaManchas have very short ears, they are tattooed on the tail web.

According to American Dairy Goat Association rules, you must tattoo a goat before applying for ADGA registration or recordation. And even if you do not have registered goats, you should still tattoo all your goats since the tattoo is a handy identification if the goat is lost or stolen.

To tattoo your goats for ADGA registration or recordation, you must use your ADGA-assigned set of unique tattoo letters. Contact ADGA or other associations for your assigned letters. (Currently there is no charge for this service.) Tattoo this set of unique tattoo letters in each goat's right ear or on a LaMancha's right tail web.

In the left ear or left tail web, tattoo a birth year letter ("Z" for 2009, "A" for 2010, etc.) plus a number. In 2009, we will tattoo kids "Z1," "Z2," and "Z3" as they are born. The letters G, I, O, Q, and U are not used because they may be confused with other marks.

For animals registered in Canada, the system is different. The letters are "W" for 2009, "X" for 2010, etc. The letters I, O, Q, and V are not used.

ADGA Tattoo Letters by Year		Canadian-Registered Tattoo Letters by Year	
2009	Z	2009	W
2010	A	2010	X
2011	B	2011	Y
2012	C	2012	Z
2013	D	2013	A
2014	E	2014	B
2015	F	2015	C
2016	H	2016	D
2017	J	2017	E
2018	K	2018	F

Tattooing Dairy Goats

Remember: if you are going to register or record a goat with the American Dairy Goat Association, its tattoo must match the tattoo information on its registration application.

If you must retattoo an animal, be sure to refer to ADGA's Tattoo Policy before you do it. (See the ADGA *Guidebook* for this.) Having an incorrect or illegible tattoo may make a goat ineligible for certain ADGA programs.

If you are registering goats with other registries, be sure to check with them (all of these associations have web sites full of useful information) for their specific tattoo requirements. Also, some registries allow microchips to be used as animal identification.

If you are raising goats other than dairy goats, you should check with your registry association to determine the requirements for permanent identification of your goats. Pygmies and Boer goats are now routinely being tattooed for identification purposes.

Breeders of hair and unregistered meat goats tend to use ear tags to identify their animals. If you do not plan on registering your goats, you should still give them permanent identification so you can recover them in case they are stolen or just wander off. Tattooing is the easiest, most effective method for doing this.

Tattooing: How To Do It

The tattooing process is easy. Here is how to do it.

Clean the ear or tail web with alcohol. Let it dry.

Using your tattoo tongs, puncture the ear firmly. Then remove the tongs from the ear. You may have to pry the ear off the tattoo needles if the ear is especially fleshy.

Put ink on the tattoo, and rub the ink into the punctures with a pencil eraser, the back of a toothbrush, or your finger. If the ear or tail web bleeds, be sure to stop the bleeding before you apply the ink, as the blood welling up will tend to wash out the ink, causing a poor quality tattoo.

Some breeders put paste ink on the ear first before they puncture the ear. They then tattoo the ear or tail and finally rub in the ink with their finger, etc. Breeders who tattoo this way say the mess on the tattoo tongs is worth it, given the excellent quality tattoos they get. However, you may find that if your tongs use metal digits, the digits may rust more quickly using this method.

We use green ink because it shows up on dark skin. We get good results with both the liquid roll-on as well as the paste tattoo ink. We do not recommend using the black ink that comes with the tattoo outfits. Black tattoos don't show up as well as those made with green ink.

Some breeders use white ink for animals with especially dark ears. They say that the tattoos show up well and do not fade.

To check a tattoo, hold a light up behind the ear you are trying to read. If you can not read the tattoo (after all, you know what you are looking for), someone else will probably have even more difficulty. Be sure you check the tattoos on all goats before entering shows or enrolling your goats in official programs.

Choosing a Tattoo Outfit

We have tried many tattoo outfits, and the ones made by Stone are outstanding. Their equipment is top quality and lasts a long time. The digits on these tattoo outfits are tab-into-slot design, which means you can not insert a digit upside down, and there is less chance of dropping one when you are changing the digits. The digits lock firmly into place, and you do not need blanks for empty spaces. The needles are nickel-plated precision-made for uniform penetration.

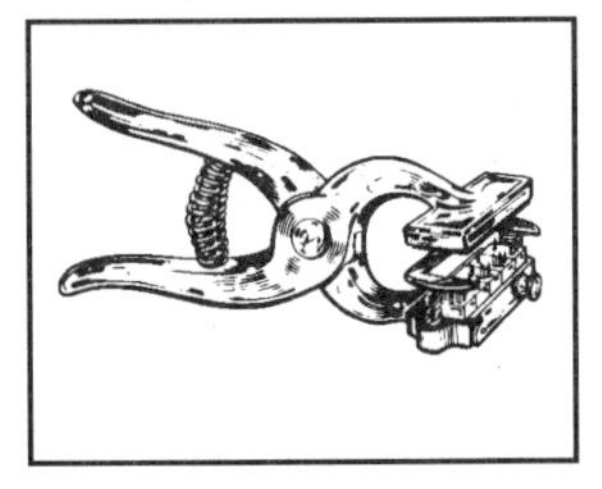

5/16 Tattoo tong

.300 tattoo tong

Both 5/16 and .300 tattoo outfits make small tattoos, much smaller than standard cattle-sized tattoos. These standard cattle-sized tattoo outfits (including all with revolving heads) are made for 3/8" digits. These digits are too large for the ears on any goats but Nubians. Even if you buy 1/4" digits to fit the big tongs, the tattoo marks are still spaced wide apart.

0 1 3 4 5

5/16 tattoo

0 1 3 4

.300 tattoo

If you raise Nubians, frequently tattoo adult goats, or want to tattoo cattle also, you may find that the 3/8" tattoo outfit will work best.

If you raise LaManchas or if most of your work is with newborn Swiss breed kids, you will like the .300 tattoo best. It is the most compact.

If you raise Pygmy goats, you will want to use the .300 tattoo outfit and probably wait a while for the kid's ears to grow a bit before trying to tattoo the tiny ear area.

Keeping Goats Healthy: An Overview

In one way goats are no different than humans: to have truly healthy goats, you need to practice preventive medicine. Just as most people automatically vaccinate their children against common diseases, do not go into areas where diseases are rampant, and ensure that their living conditions are healthful, for healthy goats, we need to do the same.

To stay healthy, goats need a clean, safe, disease-free environment. They need a quality diet appropriate to their age, condition, and level of "work." They must have copious quantities of clean water.

While goats are hardy animals, they do need some basic protection against common diseases. Most goat owners vaccinate their goats as kids and yearly thereafter against tetanus and enterotoxemia. Where abscesses are a problem, vaccination against caseous lymphadenitis may be appropriate. In selenium deficient areas, selenium supplement shots may be routine. This is a prescription substance and usually comes combined with vitamin E.

Your veterinarian is your best source of information for keeping your goats healthy. If she or he does not have specific experience with goats, ask to be referred to someone who does. In most cases a local veterinary school will have clinics and veterinarians who can be helpful, especially for difficult or unusual cases.

Recordkeeping plays an important role in keeping goats healthy. Use your barn calendar to note any health problems, observations, treatments, vaccinations administered or needed. Good records may help you uncover a health mystery and will serve as a guide for future treatments, helping you to sort out those that work from those that do not.

The key to keeping your goats healthy is you. You must take an active part in the life of your herd, checking animals each time you visit the barn to catch a problem before it turns serious. If goats are out in the pasture most of the time, scan the herd often to spot that doe who always lies under a tree or does not move with the herd or acts abnormally in some way. While kids may have different personalities (you will occasionally have a shy one), they should be active and inquisitive. Standoffishness in kids or adults may signal the beginnings of a health problem. Jump on it immediately.

Keeping Goats Healthy: An Overview

Keeping goats healthy means keeping on top of any present or potential problems.

A word of warning about the information in this chapter. All information here should be used as a guide only. While it comes from experienced veterinarians and long-time breeders, every herd is different and every goat is an individual. You should always consult with your veterinarian if you have questions about your goats' health, a specific treatment, or preventive care, or a suggestion either from this book or other goat owners you might talk with.

Remember that you and your veterinarian are in the best position to make choices about how to treat health problems in your herd. You should consider this warning also every time you read a goat health book, an article in a goat magazine or on a website on the Internet. No one knows your goats like you do, and no one treatment will fit every goat. Use resources in the way they are intended: as a guide that you and your veterinarian consider, along with the experience that you both bring to any health problem.

General Health Information

Goat Temperature? 102°-104°F

Most goats' normal temperature is between 102°F to 104°F. If you have a sick goat, the first thing you should do is take its temperature. If the temperature is above normal, there is probably an infection, either from a wound or general infection. Antibiotics might help. Below normal temperature could result from anemia, coccidia, or a mineral imbalance. However, a subnormal temperature, especially in a goat that is already sick, may indicate that the animal is critically ill.

Be sure you take your goat's temperature before calling the veterinarian for advice about a sick goat. Your vet will probably first ask what the goat's temperature is.

Routine Vaccinations for Goats

Most goat breeders routinely vaccinate their animals against tetanus and enterotoxemia.

Tetanus antitoxin provides short-term, temporary protection against tetanus. Kids should have a tetanus shot before or right after being disbudded or tattooed.

Tetanus toxoid provides long-term immunity against tetanus. Use it to vaccinate all kids and for annual booster shots for adults. For goats injured with open wounds, contact your vet about the need for further tetanus protection, even if your goat has been vaccinated annually. Also, if you have ever had horses on the property where you keep your goats, it is possible that there is tetanus in the soil. In this case, vaccination should be routine.

Enterotoxemia (Clostridium perfringens Types C & D) vaccine can be given to kids and adults. If the dam has been routinely vaccinated, the kid may be protected if raised on the dam's colostrum. Then vaccinate the kid at 4-6 weeks and again 4 weeks later. If the dam has not been vaccinated, you can vaccinate the kid the day it is born, then at 4, 8, and 12 weeks of age. Vaccinate does with a booster shot 4-6 weeks before kidding, and do not forget to vaccinate bucks. Also be aware that this vaccine is not always effective. Occasionally, a vaccinated animal will come down with the disease. Unfortunately, treatment for enterotoxemia is not very successful.

Combination vaccinations are now available, including Clostridium perfringens Types C & D combined with tetanus toxoid, and Clostridium perfringens Type D (overeating disease) combined with tetanus toxoid, and caseous lymphadenitis.

Selenium shots (Bo-Se) may be given routinely in herds located in an area where this mineral is deficient. Again, this is a prescription substance and usually comes combined with vitamin E.

Just as some people supplement their diets with vitamins, goat breeders may routine inject their goats with vitamins A and D.

If goats have problems with contagious abscesses, there is now a commercial vaccine for caseous lymphadenitis abscesses associated with Corynebacterium pseudotuberculosis. It is available from some goat supply and pharmaceutical companies. It also comes in a combined form with Clostridium perfringens Type D and tetanus toxoid. An autogenous vaccine can be prepared from material collected from your herd. It can help control abscess problems and seems to work best if the animals are vaccinated every four months.

Chlamydia can cause abortions, arthritis, and pneumonia in goat herds. Some breeders have used a chlamydia vaccine with good results. Other breeders swear by a corynebacterium pasteurella vaccine to stop respiratory and diarrhea problems in their kids.

Your veterinarian may suggest other vaccinations (such as leptospirosis) which you should use because of specific problems with goats or other livestock in your area.

Read the label carefully on any vaccine or medication. Some vaccines are more concentrated than others, so follow the label dosage exactly. If the instructions say "subcutaneous" and you give the injection intramuscularly, you may cause a severe reaction.

Injection Sites

(The accompanying drawing gives the general location of injection sites. Have your veterinarian show you exact locations and techniques.)

A *subcutaneous* injection is injected under the skin. The easiest way to give a subcutaneous injection is to pinch and lift up the skin. Then, stick the needle into the space under the "tent" of raised skin. The most common error is to inject into the skin itself rather than completely under the skin layer. The injected material ends up in the skin, and you may see swelling and a thickening of the skin at the site of injection. Most people give subcutaneous injections in the neck or behind the shoulder area.

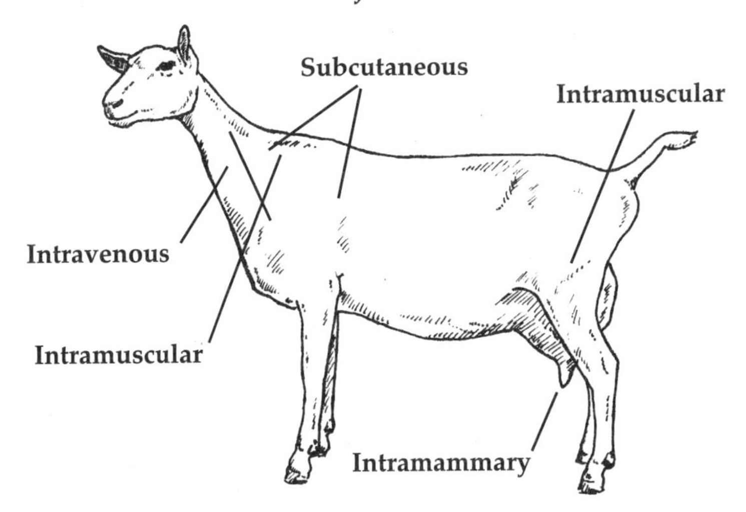

An *intramuscular* injection is given, as the name implies, deep in a major muscle like the shoulder or lower hip area. To be sure you have not hit a blood vessel, once the needle is in the muscle, withdraw the plunger a little to make sure no blood rises into the syringe. If it does, withdraw the needle and try again.

An *intravenous* injected goes into a neck or jugular vein. This type of injection is best left to your veterinarian.

Intramammary injections are given through the teat opening. Make sure that the teat is scrupulously clean and swabbed with alcohol and that the equipment used is sterile. Mastitis may result from unclean conditions. Again, your veterinarian should show you how to do this injection correctly.

Drenching

"Drenching" means to force the animal to swallow a liquid, such as propylene glycol or even water. A kitchen or turkey baster makes a very good drenching tool for goats. Measure the dose of medicine into a cup and draw the liquid up into the baster. Straddle the goat, hold its head firmly with its nose tipped up slightly. Insert the drenching tool in the side of the mouth over the tongue. Then, squeezing on the bulb, administer the liquid on to the back of the tongue. The animal should swallow the liquid.

Diseases and Common Health Problems

Brucellosis and Tuberculosis

Brucellosis (B. melitensis), which causes abortion in goats and Malta fever and headaches in humans is rare. Few cases of this infection in goats have occurred in the United States since 1972, although the disease is known in cattle, hogs, and even dogs. Scattered cases of brucellosis, allegedly caused by consuming raw milk or milk products, have been reported in Latin America.

Tuberculosis is all but unknown in goats. Testing is still recommended in areas which are not TB-free and for entering certain U.S. states, but this disease is not usually a goat health problem. Just to be safe, most goat owners test for TB and brucellosis regularly, especially if the milk is to be used for human consumption. Goats living close to the Mexican border have a greater chance of contracting these diseases because of the shipment of goats across the border.

Pneumonia

The cause for pneumonia can be bacteria, viruses, or even parasites such as lungworms. Goats are very susceptible to pneumonia and respiratory problems. They need shelter from rain and protection from drafts, but the wrong kind of shelter can be bad. Barns that are poorly ventilated, with a strong ammonia odor in the air and damp bedding, are unhealthy for goats. The viruses that cause pneumonia spread rapidly in such a setting.

Treat pneumonia as a serious disease that needs immediate treatment. Prolonged or numerous bouts of pneumonia can cause permanent lung damage undermining your goat's overall health.

Mastitis

Mastitis is an infection of the mammary system. This infection may be systemic in origin or mechanical (caused by organisms that find their way into the udder from outside). No one organism is responsible for mastitis in goats. Most udder infusions will cover a range of suspected organisms.

Subclinical mastitis is common in dairy goats, infecting sometimes sixty percent of the does in some herds. Uneven udders may indicate an infection in the smaller half. If more than a few milkers in your herd show signs of uneven udders, you may have a herd mastitis problem that needs attention.

The California Mastitis Test (CMT)

The CMT is the best method available for screening your milking herd for mastitis. The CMT reagent reacts with leucocyte cells in milk to form a gel. Some mastitis screening tests do not work well for goat milk because they do not distinguish leucocytes, a sign of infection or irritation, from normal epithelial cells, which goat milk may have in large numbers. The CMT distinguishes between these cell types. Using the CMT at least once a month cuts down on chances of early stages of mastitis going unnoticed. For diagnosis of the cause of infection, you will need to have a sample analyzed by a laboratory.

The CMT is easy to use, but the interpretation information packed with the test is for cows not goats. Research workers suggest the following interpretation of the CMT when you test goat milk:

A "trace" or "1" reaction, when the mixture looks a bit slimy, is not significant. Normal goat milk can produce this, so you can ignore this reaction. If the mixture forms a distinct gel, the "2" reaction on some CMT charts, the goat probably has mastitis. Further laboratory work should be done to determine what organisms have infected that half of the udder.

Abnormal Milk

Never use milk that is thick or stringy. If the udder is infected, the milk is not safe for humans and probably not safe for feeding kids either.

Blood specks or pink milk usually does not signify an infection. You may see pink milk when a doe has just freshened, has increased rapidly in production, or most commonly has bruised her udder, causing small capillaries in the udder to rupture.

Causes of Abortion or Miscarriage in Goats

Abortions are common in some goat herds. They are usually the result of an infectious organism such as chlamydia that causes many first-freshening does to abort or give birth prematurely, while older does are immune. Salmonella, toxoplasmosis, vibriosis, and other organisms have also been suspected in goat abortions. Severe butting, which may happen when a new doe is introduced into a herd, can also cause abortions.

Ketosis

Ketosis is a metabolic ailment that kills pregnant does. It can hit suddenly during the last month of pregnancy and kill her in a day or two if you do not recognize the symptoms and start treating her immediately. A common symptom is a wobbly or staggering gait. You can use urine tests strips to identify ketones, confirming ketosis.

Treatment is simple: administering a few ounces of sugars a day (white or brown sugar, corn syrup, molasses, and honey have all been used) will usually save the doe. Propylene glycol, a fast-acting chemical sugar, is the most effective treatment. Consult your veterinarian for appropriate dosage.

Any time a doe in late pregnancy stops eating normally, suspect ketosis. If her temperature is normal and there is no obvious infection, ketosis is a strong possibility. If you do not treat her, she may "go down," and then she may soon die. It is hard to save a doe once she is down or in the late stages of ketosis. Treating a doe you suspect of ketosis will not hurt her, even if she does not have it.

Ketosis can also hit a doe after she freshens, when she is coming into very heavy milk production. This condition usually is not fatal. It can be the cause of poor appetite, sudden drops in milk production, and off-flavored milk.

Milk Fever

Milk fever appears just prior to kidding. It is a result of a lack of calcium as the doe gears up for milk production. Symptoms may include a staggering or wobbly gait, or the doe may just "go down." While experienced goat owners may be able to treat milk fever themselves, most goat owners should contact their veterinarian at the first hint of milk fever. Feeding a low calcium diet before kidding may allow does to convert calcium in their bones more readily to calcium available for milk production.

Abscesses

Contagious abscesses caused by Corynebacterium ovis are spread by contact with pus from a draining abscess. If the bacteria get into the goat's lymph nodes, new abscesses may develop for months or years to come. Abscesses can also grow on internal organs and eventually kill the goat. External abscesses are ugly, but the goat may stay in good health otherwise. Occasional abscesses may develop inside the udder; milk from these goats should not be used for humans.

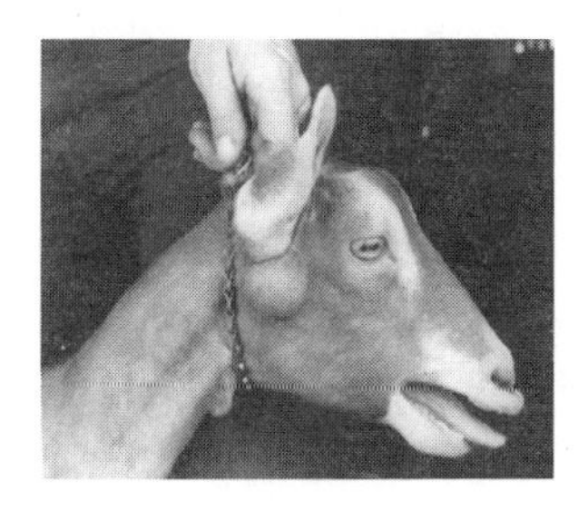

There is no cure for these abscesses other than culling affected animals from your herd. However, there is now a vaccination available to protect animals if exposed to abscesses caused by corynebacterium pseudotuberculosis. According to the manufacturer, the vaccine is of little or no benefit to animals that already have abscesses or those exposed before vaccination. Goats should be vaccinated twice at a four week interval; then they should receive a booster shot annually.

A good program of cleaning the ripe abscess and isolating the goat can reduce the incidence of abscesses in the herd. Autogenous vaccines have worked well for some breeders, but may perform best if the animals are vaccinated three or four times a year.

Abscesses may also be caused by imbedded foreign particles or small cuts infected with Staphlococcus, C. pyogenes, or other agents. These abscesses are not a contagious herd problem.

Soremouth

This highly contagious disease causes ugly sores on the mouth area of goats. The scabs themselves are the most likely means of transmission, so do not let infected kids nurse their dams. Their udders may get infected, with painful results. You may also find scabs on ears, above the hooves, teats, scrotum, and vulva.

The bad news: once one kid gets soremouth, usually every kid will eventually come down with it. The good news: one bout of soremouth usually gives lifetime immunity.

Proper treatment is to make sure goats keep eating and watch for any secondary infection of the sores.

Vaccination is usually not recommended unless you actually have the disease in your herd because the vaccine is "live" (it will infect your premises). A vaccination program (when followed rigorously) has helped clean up some herds with outbreaks of soremouth.

If you choose to apply an antibiotic ointment, be sure to protect yourself by wearing gloves. If the virus gets into a cut on your hand, you too will probably get soremouth.

"Off Feed" and Teeth Grinding

Eating should be the primary interest in any goat's life. If the goat stops eating normally, the animal is probably sick. Going "off feed" is one of the few ways a goat can tell you if it does not feel well.

Teeth grinding is also a sign of illness in goats. You can easily hear this unpleasant noise and can tell that the goat is uncomfortable and needs your attention.

Ringworm

Roundish hairless patches on a goat's body are often caused by a ringworm fungus. It can be more a bother than a health problem. Older does will usually get ringworm only if they have never been exposed to it as kids, while virtually every kid may get at least one spot, and many may end up covered with lesions.

Goats usually are not seriously ill from ringworm. As with soremouth, make sure that the lesions on affected goats do not get infected.

People have tried various treatments with varied success: Kopertox, wormer paste, equal parts of glycerin and tincture of iodine, household bleach diluted 1 part in 10 parts water used daily may be of some use, but nothing seems to work as well as the Fungisan Liquid. It helps prevent or at least delay the onset of ringworm in unaffected animals, and helps clear up lesions on affected animals. To use it, clean the skin with a mild antiseptic soap and then apply.

It is important to remember that if you take infected animals to a show or another farm, you are exposing all animals your goats come in contact with to ringworm. So be a good neighbor and keep goats with ringworm home until the ringworm is cleared up completely.

Pinkeye

This contagious eye infection, caused by bacteria, viruses, or chlamydia, among others, may spread through a whole herd. Severe cases of pinkeye may cause temporary blindness, so be sure affected animals eat regularly. Treat affected goats with an antibiotic such as Terramycin eye ointment. Since pinkeye is so contagious, you may decide to treat not only affected goats but all goats at least once at the first sign of pinkeye in a herd.

Lice

If a goat scratches quite a bit, it probably has lice. Some goats are more susceptible to lice than others, and lice can be serious skin parasites. Often the single emaciated kid in a pen will have a huge load of body lice.

To keep lice under control, use a good dairy insecticide powder. It is also worthwhile to clip goats completely in warm weather. This helps get rid of lice. However, do not powder the barn cat. Coat-licking may cause a fatal ingestion of insecticide.

Treating Scours or Diarrhea in Kids

Experience has taught us to treat all cases of scours in kids as potentially serious. This means that you should treat kids at the first sign of diarrhea. Too often, a kid with mild scours can, for one reason or another, go downhill quickly either due to dehydration or other causes, so quick action can avert potential disaster.

For mild scours, give kids Kaopectate. You can use a 6 cc syringe without needle as a drenching tool. Reduce the amount of milk per feeding and feed more frequently.

For severe scours, stop feeding milk completely, and replace it with an electrolyte solution. This will help replace body fluids lost through dehydration (the real problem when kids have severe scours), and its pre-digested protein is an excellent nutrient source. Since most electrolyte solutions have no medication, treat affected kids with Kaopectate or other scour medicine also. When the scours stop (usually within 12 hours on this regime), do not start feeding whole milk immediately. Start by feeding half milk, half electrolyte solution, in small amounts, several times a day.

If diarrhea has not stopped in 24 hours, begin treating with an oral neomycin also. If you have any doubts about your kid's condition, consult your veterinarian. Do this sooner, rather than later. Scours can kill kids, or set back their growth seriously.

To check for dehydration, pinch the kid's skin. If the skin sticks together, the kid is seriously dehydrated and should be treated with subcutaneous injections of dextrose-saline solution, 50 ml. total. Again, check with your vet if you have any questions about dehydration and how to treat it.

Scours or Diarrhea in Kids

If you do not have an electrolyte solution on hand, you might try canned beef broth. Most important, stop feeding milk and replace it with some high nutrient solution.

If scours persist in a single kid or you have scours in a number of kids, take fecal samples to your vet to be checked for coccidiosis. Coccidiosis can be treated with Amprolium or with various sulfas (see the chart, "Drugs for Controlling Coccidia" for more specific drug information). Medication is usually given in drinking water or in the kid's milk.

Many scour problems in kids are mechanical (too much food eaten too fast). Monitor your kids' eating more closely, and you can usually eliminate this type of scours. Also, dirty bedding and overcrowding in kid pens can set off scours in kids.

Emergencies—What YOU Can Do

While most accidents or health problems are minor, occasionally something disastrous happens, and we are faced with a real emergency. Most of us know a little first aid, but it is reassuring to have a short guide to follow when disaster strikes. Here is what our veterinarian suggests as practical emergency care.

For Cuts or Lacerations:

If there is a large amount of bleeding—to the point of weakness, panting or coma—FORGET about cleaning the wound. You first need to control or stop the bleeding.

A kitchen towel soaked in ice water will make a good temporary bandage. Either hold it on the wound with your hands, applying moderate pressure, or if you need to leave the animal for a moment, wrap the bandage securely with vetwrap, duct tape, rope, anything that will keep the bandage in place and some pressure on the wound.

In cases of severe bleeding, seek veterinary assistance immediately. Keep the animal warm and quiet until help arrives. You might keep an old quilt or blanket handy for just such an emergency.

For less severe cuts, use clean, warm soapy water (surgical soap works well here) and a clean cloth to bathe the wound. Be sure to wash away any dirt particles, dried blood, pus, or other signs of infection. Bathing the wound may help cut down on any bleeding.

At this point you will need to decide whether the wound requires stitches to close it. Generally, a gaping wound that looks as if it is not likely to close and heal naturally may require a veterinarian's assistance. Otherwise, if no stitches are needed, apply a topical antibiotic and, if the wound requires it, apply a bandage secured with tape.

Be sure not to bind the wound too tightly, and check it daily for signs of infection (the signs include inappropriate swelling, redness, and/or pus).

There are two ways to approach the issue of bandages. Some people like to keep a wound open to the air for quick healing, so they do not use a bandage or remove it as early as possible. Others believe that since goats live in a pretty "dirty" world, a bandage protects the wound until a scab forms and healing is well on the way.

For Broken Limbs (treatment depends on the type of injury):

If you are not sure that the limb is fractured (there is no obvious broken bone), first try to immobilize the joint above the suspected break. You can wrap a heavy towel around the leg and hold it fairly tight with vetwrap or duct tape. You can also use wood splints wrapped with a towel or cotton (wood shingles cut to size work well) or rolled newspapers or magazines. Hold these together with vetwrap or duct tape. Seek medical assistance.

If there is an obvious fracture, do not attempt to set the bone yourself. First, try to immobilize the animal, keeping it from thrashing around. If the bone has broken through the skin, coat the wound and bone with an antibiotic, cover the bone and wound with a clean towel or bandage, and then try to immobilize the joint as suggested above. Seek medical attention for the injury immediately because this type of injury to the bone can very easily lead to infection. As with all serious injuries, keep the animal warm with blankets if it looks the least bit weak or in shock.

A bit of advice: our veterinarian says that horses cause more broken bones in goats than any other cause. So if you keep goats in with horses, monitor your goats carefully, and move them out at the first sign of trouble.

If you must move the goat, make sure you immobilize the limb as best you can. If you are alone, phone a neighbor for help. Otherwise, try to get your vet to send someone who can at least keep the animal from thrashing around while you drive. It is very difficult to transport a seriously injured animal if you have to drive as well as keep the animal calm.

For weak kids

If you have a weak newborn that will not nurse, try the following. Dilute some Karo Syrup in 2 pints of warm water and carefully dribble a little into the kid's mouth. Do not feed the baby too much or too fast. A Pritchard teat works best for this type of feeding as it delivers just a little bit of the solution into the kid's mouth. This solution supplies quick, digestible energy, usually enough to get the kid strong enough to begin normal nursing. If you are desperate and do not have any Karo syrup, you can try feeding a tiny bit of black coffee. It will often provide enough stimulant to get the kid going. Be sure to follow up any solution you feed with colostrum. The syrup or coffee is just meant to give the kid a "kick start."

If the kid is too weak to nurse, you might try tube-feeding it, using a weak kid syringe. If the kid fails to respond quickly, seek medical attention.

Other Emergencies

If you find an animal in obvious distress, first calmly but quickly survey its vital signs and general condition so you can give the veterinarian the best opportunity to help you.

Check the rate of breathing. Is it rapid, slow, labored, shallow? Does the goat show signs of bloat? Normal rumen movement is about 1 to 1.5 movements per minute. Can the animal move its limbs or are they rigid? Outstretched?

Is there any mucous around the nose, mouth or eyes? Any bloody discharge from the mouth, nose, vagina, rectum?

Any fever? Low temperature? Is the animal in shock? Normal respiration is between 12 and 15 per minute. The kid rate is faster. A goat's heart will beat approximately 70 to 80 times per minute, a kid's rate is again faster. A healthy animal usually has pink, rosy gums; a shocky one has pale gums. Is the animal dehydrated? Pinch a bit of skin on the neck. If it does not snap back fairly quickly, the goat is probably dehydrated.

In pregnant does, smell the breath or urine; if it smells like acetone, she may be ketotic.

While this is not a complete list, checking these aspects of the animal's condition before calling the vet will help him or her give you the best advice quickly.

Portions of this article originally appeared in the April/May, 1998 Goat Magazine. *We wish to thank both the author, Gail Bowman, a Boer Goat breeder from Twin Falls, Idaho, and the publisher, Roylyn Coufal, for allowing us to excerpt from this article. You'll find a list of laboratories that test for CAE in the Resources section of this book.*

CAEV

by Gail Bowman

Caprine arthritis-encephalitis virus (CAEV or what is often abbreviated to CAE) is a common disease of goats, prevalent worldwide. In the United States, an 81% prevalence rate in goat herds has been reported. According to a recent study, "Infection most commonly occurs when the virus is present in colostrum or milk that is ingested. However, prolonged contact, particularly in high-density goat populations, also results in significant transmission" (Rowe and East, 35).

Sources of Infection

Not all CAEV infections in kids can be explained by ingestion of affected colostrum or milk. Up to 10% of kids who were removed from their "positive" mothers (mothers who have tested positive for the CAE virus) have been reported to show infection. In these cases, researchers posit that infection must have occurred in one of four ways: in-utero transmission, transmission from the dam by vaginal contact, accidental ingestion of infected colostrum, or transmission from the dam by exposure to saliva or respiratory secretions during licking.

Symptoms of CAEV

CAEV infection may not be serologically (blood) testable for months or years, and some infected animals who can transmit the disease may never show clinical symptoms. Symptoms include a progressive rear leg weakness and/or paralysis in kids two to six months of age, chronic arthritis in adults (most frequently found in, but not solely restricted to the knee joint), inflammation of the mammary gland, lung, and nervous system. Nervous system involvement may include blindness, head tilts, and facial nerve paresis. Mammary involvement results in udder edema or "hard udder," where the entire udder becomes hard and warm within the first few hours after kidding. (East, 591-600) Lung involvement results in chronic pneumonia.

Sources of Transmission

In milking herds, shared milking machines, milk contaminated hands or towels, etc., will significantly increase the risk of spreading the disease. In meat herds, transmission may occur via needles, tattooing instruments, or dehorning equipment. Also, in high density herds that are endemically infected, head butting to the point of drawing blood, eye-licking, biting, snorting and coughing, and urinating too near another goat's face may all be possible sources of infection (Greenwood, 341). In addition, according to Rowe and East, research shows indications that sexual contact may bring the risk of infection through the exchange of saliva, estrus mucus, urine, semen, and nasal secretions (41).

Negative to Positive Conversions

Researchers estimate that the period of time between exposure to CAEV and development of detectable (testable) antibody levels to be between three weeks to eight months after exposure to the disease (Oliver, 158; Ellis, 242; and Rimstad, 345). Many breeders have reported conversion from negative test results to positive test results in goats as old as five years of age, although conversion seems to be most prevalent between one and two years of age. Researchers attribute conversions later than two years of age to the lateral exposure to the disease as discussed above.

To Help Prevent CAEV Transmission

Veterinarians recommend taking the following steps to help prevent CAEV transmission:

- Immediately remove the kids from their dams at birth. Try to keep the birth sack intact until the kid is out of the mother's body. Do not allow does to lick their newborn kids. Wash newborns in warm water in a clean sink.
- Feed newborns only heat-treated, artificial, or cow colostrum. Do not pool colostrum from mothers that might be infected and then feed the pooled colostrum to kids.
- Feed only pasteurized milk or goat milk replacer (never raw goat milk) to kids until they are weaned.
- Separate all possibly infected animals from uninfected animals by a double fence with at least ten feet between the fences. Do not use common feeders, waterers, or salt blocks.
- Milk negative and younger does before milking positive and older does.

- When possible, breed negative does with negative bucks. If negative and positive animals are mated, use a single hand-mating, allowing minimal oral contact.
- Do not share needles, tattooing equipment, or dehorning devices without careful cleaning and sterilization.
- Test kids at six-month intervals starting after the kids are at least four months old. Most breeders suggest testing during times of stress, such as about two to three weeks before kidding. There is, however, some evidence to suggest that a goat with another systemic infection may test with a false positive for CAEV (Rowe).

Testing for CAEV

You can easily and inexpensively test for the CAEV by drawing your own blood samples and sending them to a laboratory for testing. Use a 3cc syringe and an empty, sterile, red top blood vial. Label the vial as the laboratory instructs, and send it by Next Day delivery service to a good lab. A list of laboratories is provided in the Resource section of this book.

References

Rowe, Joan Dean, and Nancy E. East. "Risk Factors for Transmission and Methods for Control of Caprine Arthritis-Encephalitis Virus Infection." *The Veterinary Clinics of North America,* Mar 1997, Vol. 13, No. 1.

East, Nancy E. "Diseases of the Udder." *The Veterinary Clinics of North America,* Nov 1983, Vol. 5, No. 3, 591-600.

Greenwood, P.L., R.N. North, and P.D. Kirkland. "Prevalence, spread and control of Caprine Arthritis-Encephalitis Virus in dairy goats in herds in New South Wales." *Aust Vet J* 72:341, 1995.

Oliver, R.E, R.A. McNiven, and A.F. Julian, et al. "Experimental infection of sheep and goats with Caprine Arthritis-Encephalitis Virus." *NZ Vet* 30:158, 1982.

Ellis T.M., H. Carman and W.F. Robinson, et al. "The effect of colostrum-derived antibody on neo-natal transmission of Caprine Arthritis-Encephalitis Virus infection." *Aust Vet J* 63:242, 1986.

Rimstad E., N. East, and E. DeRock, et al. "Detection of antibody to Caprine Arthritis-Encephalitis Virus using recombinant gag proteins." *Arch Virol* 134:345, 1994.

"We found this dosage conversion chart very handy for treating our goats."

1 ml	=	15 drops	=	1 cc
1 Tsp	=	1 gram	=	5 cc's
1 Tbsp	=	1/2 oz	=	15 cc's
2 Tbsp	=	1 oz	=	30 cc's
1 pint	=	16 oz	=	480 cc's

"You can pick out healthy goats a mile away. They have a liveliness and bounce as they walk. Their coats shine, their eyes twinkle, and they seem to take pleasure in their lives."

"Check all wounds daily to catch the first signs of infection, including redness, puffiness, and weeping or oozing."

"As in humans, it's really important to finish the course of any medications that your veterinarian prescribes for your goats. If you don't finish the medicine, your animal may relapse or have other problems. Even if the goat looks good and acts lively, continue the treatment to the end."

"Be sure to read all label directions to make sure you understand how to administer a medicine and how long to give it. If you have any questions, be sure to clear them up with your vet."

"Take your goat's temperature before calling the vet. The first question most vets ask is, "Does your goat have a fever?"

"When an animal of any kind has a wound, its natural instinct is to lick it. This often will keep the wound from healing and cause further health problems. One way to "lick" this problem is to use what vets call a "Victorian collar," named for the massive ruffs people wore during the reign of Queen Victoria, I surmise. Take a small bucket, cut out the bottom and fit it over the goat's head. The bucket should rest below its ears and keep the goat from being able to move its head around to lick. The trick is to find just the right size bucket for your goat. You can also ask your vet about these because some of them stock this type of commercial collar designed for house pets.

"Keep a clean leg snare in a plastic baggie in your barn in case you need it quickly to save a kid that's having trouble being born."

Health: Hints

"We almost lost a nice older doe to coccidiosis. She never had the typical scouring that you see in kids. Her only symptom was that she lost weight and was listless. After I treated her with Amprolium, she started to gain back weight and looked and felt a whole lot better."

"After fighting coccidiosis in our kids for years, we now routinely worm our babies against coccidiosis at 21 days and then every 21 days until they're pretty well grown. With this program, we haven't had any coccidiosis problems and our kids are big and strong."

"Use your barn calendar to keep track of all treatments and health problems."

"To make sure all our does get wormed in the spring, we worm them at the same time we milk out their colostrum for the first time. That way we know we've gotten around to all our fresh does."

"Put some udder cream on the udder of a doe that's ready to kid. It helps relieve udder congestion and helps keep bedding and birth fluids from sticking to the udder."

"Udder cream isn't just for udders. We use udder cream on goats the same way we use hand creams on us. If we find a rough spot on the skin or some chafing, we put udder cream on it."

"Most pharmaceutical products are approved for cattle, sheep, or hogs, not goats. The reason for this is that it takes costly testing to get FDA approval to label a product for a species, and there has not, in most cases, been enough potential profit from goatkeepers to justify the investment."

This article and chart were originally published in the Winter 1997 issue of Goat Tracks. *We wish to thank Ellen Herman, publisher, and Nat Adams, DVM,* author, for allowing us to reprint this excellent work here.*

Drugs for Parasite Control

by Nat Adams, DVM, Denton, TX

The information in the chart is subject to change at any time. It is designed to give goat owners a concise source from which to make logical decisions on parasite control. It might be helpful to your veterinarian in working out a successful worming program to fit your herd. In order to be effective, a good parasite control program must be implemented with due consideration to laboratory fecal samples taken at intervals appropriate to climate and types of parasites in your area.

Be sure the weights of your goats are accurate when figuring dose. Some of these drugs are not safe when overdosed, and some are ineffective when underdosed.

When giving injections of any medication, be sure to use sterile techniques and a separate needle for each animal. Sterile means autoclaved or new and uncontaminated. Alcohol rinse is not a proper sterilizing solution for livestock equipment, and residues of any sterilizing solution may affect the drug being dosed.

If money is a limiting factor (as it is with most of us), price these drugs according to how many doses the container provides.

My favorite combination of drugs for worming goats is:

- Ivermectin cattle injectable given **orally** at 300 micrograms per kg. for major worms.
- If I see tapeworms, I use fenbendazole, one dose at 10 mg. per kg.

The chart that follows is a composite of information from:

Compendium of Veterinary Products, 3rd ed. (1995-1996). Pub. by North American Compendium, Inc., Fort Huron, MI.

Veterinary Drug Handbook, 2nd ed., by Donald C. Plumb. Iowa State University Press, Ames, IA, 1995.

Personal Experience.

*Dr. Nat Adams has been a practicing veterinarian for 29 years and has raised Alpine dairy goats for the past 25 years. She got her start in dairy goat medicine working with Dr. Sam Guss.

Drugs for Parasite Control

Target Parasite	Brand Name	Drug	Recommended Dose	Translated Dose	Withholding Time	Notes
Adult gastro-intestinal nematodes & some of their larvae	Valbazen	Albendazole	Oral, 10 mg per kg.	450 mg per 100 lbs.	Cattle, meat 27 days. Milk, not approved	1. Do not give in first trimester of pregnancy. 2. Do not use within 2 weeks of anesthesia.
Same as above	Panacur Safeguard	Fenbendazole	Oral, 5 mg per kg.	225 mg per 100 lbs.	Cattle, meat 8 days.	1. Do not use with other medicine for flukes.
Same as above	Ivermec	Ivermectin	Subcutaneous injection, 200 micrograms per kg.	Subcutaneous injection, 1 cc per 100 lbs. Painful!	Cattle, meat 35 days. Pkg. says cattle, 49 days. Still present in milk 30 days after injection.	1. Dr. Craig (Parasitologist at Texas A&M Vet School) recommends **oral** dose of this for goats at 300 micro-grams per kg.—about 1 cc per 75 lbs. 2. Also useful in treating skin fungus, lice (sucking variety only) and skin mites. Personal experience with this is injection only and at a higher dose. 3. A cattle pour-on works well but only if goat's back is shaved. No known dosage. Try recommended dose on box. 4. If you have a heavy grub problem, give this only in the dead of winter.

Drugs for Parasite Control Continued:

Target Parasite	Brand Name	Drug	Recommended Dose	Translated Dose	Withholding Time	Notes
Some gastro-intestinal nematodes	Ripercol and Tramisol	Levamisol	Subcutaneous injection or oral tablets at 8 mg per kg.	360 mg per 100 lbs.	Cattle, meat 7 days. Sheep, 3 days	1. Do not use with other wormers, antibiotics, or insecticides (within 2 weeks). 2. Not for breeding or lactating animals. 3. Vital to follow the above cautions.
Gastro-intestinal nematodes	Rumatel	Morantel	Sheep dose: Oral, 10 mg/kg	450 mg per 100 lbs.	Cattle, meat 14 days. Milk, 0 days	1. Do not give with mineral or bentonite supplements. Do not give with other wormers or insecticides. 2. Another source says OK with insecticides. 3. OK for lactating and pregnant cows.
Same as above	Benzelmin	Oxfendazole	Horse dose: Oral, 7.5 mg per kg.	350 mg per 100 lbs.	Not for food animals.	Not for food animals.
Same as above	Anthelcide	Oxbendazole	Sheep dose: Oral, 10-20 mg per kg.	675 mg per 100 lbs.	Not for food animals.	Not for food animals.
Same as above	Imithal Strongid	Pyrantel	Horse dose: Oral, 25 mg per kg.	1125 mg per 100 lbs.	Not for food animals.	Not for food animals.
Same as above	Rintal	Pebantel	Horse dose: Oral, 5-10 mg per kg.	350 mg per 100 lbs.	Not for food animals.	Not for food animals.

Drugs for Parasite Control

Target Parasite	Brand Name	Drug	Recommended Dose	Translated Dose	Withholding Time	Notes
Gastro-intestinal nematodes	TBZ Thiabenzole Omnizole Equizole	Thiabendazole	Oral, 44-66 mg per kg.	900 mg per 100 lbs.	Goat meat, 3 days. Milk, 96 hours.	1. Only wormer on chart government approved for goats. 2. Large population of parasites resistant to this drug. 3. Very safe for pregnant and debilitated animals.
Ascarids (round worms) & Oesophag-ostomum	Numerous brands	Piperazine				Mentioned here because although it is a common drug, it has relatively little benefit.
Lungworms	Valbazen	Albendazole				See original listing above.
Lungworms	Panacur Safeguard	Fenbendazole				See original listing above.
Lungworms	Ivermec	Ivermectin				See original listing above.
Lungworms	Ripercol Tramisol	Levamisol				See original listing above.
Lungworms	Rintal	Pebantel				See original listing above.

Drugs for Parasite Control Continued:

Target Parasite	Brand Name	Drug	Recommended Dose	Translated Dose	Withholding Time	Notes
Tapeworms	Valbazen	Albendazole				See original listing above.
Tapeworms	Panacur Safeguard	Fenbendazole	10 mg per kg.	500 mg per 100 lbs.	Cattle, meat 8 days	This is double the dose used for nematodes.
Tapeworms	Drondt	Praziquantel	Dog dose: Subcutaneous injection/oral 10-15 mg per kg.	350 mg per 100 lbs.	Not for food animals.	Very expensive.
Flukes	Valbazen	Albendazole				See original listing above.
Flukes	Drondt	Praziquantel				See original listing above.
Flukes	Rintal	Pebantel				See original listing above.

This article and chart were originally published in the Spring, 1997, issue of Goat Tracks. *We wish to thank Ellen Herman, publisher, and Nat Adams, DVM,* author, for allowing us to reprint this excellent work here.*

Drugs for Controlling Coccidia

by Nat Adams, DVM, Denton, TX

Coccidia are intestinal parasites that can cause severe protein loss in goats. These are not worms. They are microscopic-sized animals (one cell only). The average worming medication will not eliminate these pests. In fact, to control coccidia, you must approach the problem with a whole different attitude. Once coccidia is in your herd, you must consider the age of the animals, weather, housing, sanitation, and immunity, as well as medication in planning your approach to controlling it.

The chart that follows will help you pick medications that will fit your plan.

For active cases of coccidia, I like to use Albon tablets daily for five days, then Corid in drinking water at the lower dose for 21 days. I never allow Bovatec on my property because of the possibility of accidental horse poisoning.

*Dr. Nat Adams has been a practicing veterinarian for 29 years and has raised Alpine dairy goats for the past 25 years. She got her start in dairy goat medicine under the tutelage of Dr. Sam Guss, during the seven years that she practiced in New England.

Drugs for Controlling Coccidia

Brand Name	Drug	Recommended Dose	Translated Dose	Withholding Time	Notes
Decox	Decoquinate	Oral, daily .5 mg/kg in feed 1/2 lb/ton	22.7 mg per 100 lbs. Feed 1 lb per goat	None, not for lactating dairy goats	1. Use as a feed additive. 2. Kills only immature coccidia, so it's good only as a preventative. If goats have active coccidia, treat also for mature coccidia.
Bovatec	Lasalocid	Sheep, oral, daily, 10-30 gm/ton of feed. Or 15-70 mg per head per day		None	1. Use only as a feed additive, not an individual dose. 2. Kills equine (horses). 3. Cannot be used in animals intended for slaughter before 4 months of age.
Albon Bactrovet	Sulfadi-methoxine	50 mg/kg 1st day. Then 25 mg/kg per day for 5-13 days.	500 mg/20 lbs 1st day. Then 250 mg/20 lbs daily for 5-13 days.	Cattle, meat, 5-7 days. Milk, 60 hours.	1. Keep animal well hydrated. 2. Liquid Albon tastes great but it's expensive.
Sulmet liquid	Sulfadi-methoxine	112.5 mg/lb 1st day. Then 56.25 mg/lb for 4 days	6 Tbs/100 lbs 1st day. Then 3 Tbs/100 for 4 days.	Cattle, meat, 10 days. Is passed in milk.	1. Use for 5 days only. Very inexpensive. 2. When used in drinking water, a 100 lb animal will drink about 1 gallon of water per day. 3. Weather and illness may cause water consumption to vary considerably. Use caution.

Drugs for Controlling Coccidia Continued

Brand Name	Drug	Recommended Dose	Translated Dose	Withholding Time	Notes
Corid	Amprolium	Treatment: 10 mg/kg per day for 5 days. Prevention: 5 mg/kg daily for 21 days	See note #4	Cattle, meat, 24 hours.	1. Feed or water additive. 2. May use as individual drench. 3. Mix fresh each day. 4. Excellent mixing instruction on pkg. 5. Thiamin (B_1) and Amprolium are antagonistic. Overdose or long-term use can result in central nervous system problems. Giving B_1 supplements during treatment will decrease effectiveness of the coccidiostat.
Rumensin	Monensin	20 gm/ton of feed		Cattle, meat, none.	1. Kills equines (horses). 2. Requires 5-day break-in period. 3. Do not feed to lactating animals. 4. Use only in feed mixture. 5. Feed as only source of grain.

Drugs for Controlling Coccidia

Showing Goats: An Overview

The first time you ever attended a county or state fair livestock show, you probably wondered just what was going on. What were the judges doing as they walked around the ring and pointed out animals? Why did the exhibitors spend long hours conditioning their animals, hauling them from place to place, enduring often primitive conditions at the fairgrounds, all for, in many cases, very little recognition and even fewer cash awards?

Livestock expositions and competitions are very popular, especially with children and young adults who are active in 4-H and FFA projects and activities. For species other than goats, there seem to be more rewards, both tangible and intangible. One can find impressive show premiums and livestock auction prices for meat animals. Young exhibitors of beef cattle, swine, and sheep have financed their higher educations from the proceeds of their livestock projects.

For those raising dairy projects, the cash awards lag far behind the beef cattle and other meat animal projects. However, we are seeing both a growing interest in showing meat goats, plus more recognition of the appeal and worth of dairy goats. We can hope that with a growing market for goat meat and milk products that exhibitors of all goats will begin to reap greater financial rewards and community support.

Showing livestock is not for everyone. Many of us are happy to breed our animals, sell or give away the occasional unwanted offspring, and never feel the need to compare our herd, in a competitive way, to anyone else's. The majority of goat owners will never exhibit at any type of goat show.

Most people get into showing goats by attending a local show. It looks like fun, and soon they may get "bitten by the show bug." Showing goats may become the center of their lives. Breeding and dry dates may be planned to correspond with specific show dates and class restrictions. Goats may be purchased to fill "holes" in a show string. Some may deny themselves day-to-day little luxuries so they can participate in a national goat show or state fair.

Why? Money is rarely the reason. While state and county fairs may pay premiums for those placing at the top of their classes, it is rare unless you are one of the very top exhibitors that your show winnings can do more than cover your expenses. Shows put on by local goat clubs rarely pay any cash awards. The most you can hope for is a nice trophy.

Depending upon where you live, you could attend a dairy goat show every weekend from April to October. There are also numerous Boer goat shows, and we see more shows for Pygmies and Nigerian Dwarfs, too. If you enter more than a few shows during the show season, your expenses for entry fees, travel, and day-to-day expenses will add up quickly. Unless you have a very small herd all of which you take with you, you will also need to arrange for someone to do your chores while you are at the show.

While it is obvious that showing does not return huge cash rewards, those who participate actively value it highly. If you are planning to sell breeding stock, especially nationally, you will probably need to have show records on your best animals. Some buyers are interested in production and production records alone, but the majority who are trying to better the conformation in their herds will want show records to help guide them.

If you think you want to try showing your goats, first attend a show to see just what goes on. Talk with people about where the upcoming shows are and what shows are appropriate for the goats you have. People who show goats are more than happy to share tips on when to arrive, how to arrange your pens, how to prepare your goats, how to set up your goats in the show ring, what papers you will need, etc.

Put your name on the mailing list of local clubs that have shows and you will receive mailings and entry forms. Visit a local breeder when he or she is clipping goats for a fair. Offer to help an exhibitor at a show. The more you learn before you actually take your goats to a show, the easier your experience will be and the more likely that you will enjoy showing, which is the whole point!

Here Comes Da Judge

Are there different kinds of shows?

For dairy goats, you can attend both local shows, usually sponsored by a dairy goat club, and shows sponsored by county or state fairs. The differences between the two vary from area to area, but usually club shows do not pay premium awards, while county and state fairs may. Club shows may have more relaxed rules and offer classes intended to accommodate novices and younger exhibitors. You may see a costume class or adult showmanship class at a local show but not usually at a fair show.

For dairy goats, the shows you attend will usually be divided into shows specifically for dry animals (those under the age of two that have never freshened) and milkers. There may also be a separate show for bucks. Sometimes at larger shows and at fairs, there may be a youth or 4-H show, specifically for children and young adults. Showmanship classes are popular at both local and fair shows.

What makes a show official?

For dairy goats in the United States, an official show is one sanctioned by an association such as the American Dairy Goat Association or the American Goat Society. The host club will apply for the sanction ahead of time, and the sanctioning association in turn sends official papers. For other types of goats, you should check with the respective associations about official show sanctions.

Who can be a judge?

To be a judge for most shows sanctioned by the various goat associations, you must have passed some kind of licensing training. This training varies from association to association, but judges usually have a number of years of experience raising and showing goats. They may have to pass a written test and be evaluated by other, more-experienced judges as they work classes of animals. They may have to pass refresher courses over the years.

The American Dairy Goat Association has a licensing program, whereby a candidate attends a pre-training conference where he or she learns about the scorecard and practices giving accurate reasons for the goats being judged. Next comes a written test, and finally the hands-on judging of classes of goats. Candidates are scored on accuracy of placings and reasons by a panel of experienced judges. Candidates can earn licenses spanning one to four years. Those holding continuous licenses over a certain time span may become senior judges and return for refresher sessions.

What is the judge looking for?

Most associations have an approved scorecard that judges use to evaluate animals. This scorecard includes drys, milkers, and bucks. Judges evaluate animals for specific traits in general appearance, dairy character, body capacity, mammary system (for milking does), and reproductive system for bucks. After having the class walk around the ring, the judge will evaluate the animals one by one and then compare them one to another. Placings usually come next, with animals lining up first to last. The judge will then give oral reasons, telling why one animal places over another.

How are champions chosen?

Goats are shown in age classes. In general, to determine the grand champion of the show, the first place winners of all age classes return to the ring, and the judge chooses the best goat overall. To choose a reserve champion dairy goat, the judge will choose from the remaining first place winners and the second place winner to the grand champion (if there is one).

Many shows will choose a best in show winner also. For this class, all grand champions (or best of breed winners) return to the ring for judging. The judge will give oral reasons for his or her choice here also.

Prizes for winning may be simply a ribbon or in some cases elaborate trophies and cash awards.

What is a permanent champion dairy goat?

Dairy goats who win a specified number of official grand championships may be awarded permanent champion status. Their papers will reflect a CH or GCH before their name. They may still show in regular classes at local or fair shows, or if available, they may show in special champion challenge classes, where they "challenge" the chosen champion for best of breed status. The best of breed winner then participates in the best of show judging.

Where do I get information about shows?

For information about upcoming shows and the rules that apply, contact the registry association for the goats you want to show.

Grooming Dairy Goats for the Show Ring

This article appeared originally in the June, 1983, United Caprine News. *We wish to thank Jeff Klein, publisher, for allowing us to reprint an edited version of it here.*

Basic Techniques: Grooming
by Jeff Klein

The official shows program of the American Dairy Goat Association gives us, as dairy goat owners and breeders, the opportunity to put our animals before the public. Certainly, this public presentation of dairy goats should be one that will gain the admiration and approval of the viewing audience. This requires that animals be well groomed.

But there is more to grooming than simply making the animal look appealing to an audience. A properly fitted dairy goat will gain in general appearance and will more readily exhibit smoothness of blending. This makes the job of the judge much easier—he or she does not have to guess what is underneath a long coat of hair.

Below are basic grooming techniques, although you may want to change some of these procedures as you gain experience.

Grooming the animal is only one small part of the entire fitting program. Experienced showmen know that an animal must be in proper condition—not too fat or too thin—to do well in the show ring. Growth and size often enter into a judge's final decision. After all, such attributes are needed (especially in kids) if they are to be large enough to breed as yearlings. Thus, fitting begins the minute a kid is born and continues until the day that the animal is retired from the show ring. Health problems and poor nutrition will hurt an animal's performance in the show ring.

Grooming here includes clipping the hair coat, trimming the hooves, and performing those final preparations to present a dairy goat properly in the show ring.

Advice on Clipping Kids

When the weather is variable, clipping a young kid may be asking for trouble. I prefer not to clip my kids if the weather is iffy. I simply trim them so that the long hair on the thighs, barrel, and around the hooves is removed. I may err on the side of caution, but a kid who gets sick after being clipped for an early show may never regain the bloom she had to begin with. You will have to use your judgment here.

Clipping Does

The first step in grooming is to clip the hair over the entire body of the goat. This is best done using a good pair of electric clippers sold for clipping horses or dogs. I use both the Sunbeam Clipmaster [consider also the Lister Star and Andis Progress clipper, ed.] and a small animal clipper like the Oster [consider Wahl clippers and others, ed.] that has interchangeable blades.

If you are just beginning, you will probably want to invest in only one set of clippers. I suggest the smaller type. Because it is smaller and lighter than the Clipmaster, it is easier to clip legs, face, and udder, and its light weight will not tire your arm as much. The Clipmaster has a larger clipping head which makes the clipping job faster, but you will need practice to perfect using it on kids and small areas of does.

If you have the small clippers only, use the standard No. 10 blade. Experienced breeders use the 30, 40, or even the 50 blade on the doe's udder just before the show. These are surgical blades and actually shave the udder. This gives the udder a bright, shiny appearance and helps make the udder look more capacious.

The diagram below shows the type of clipper I use on each part

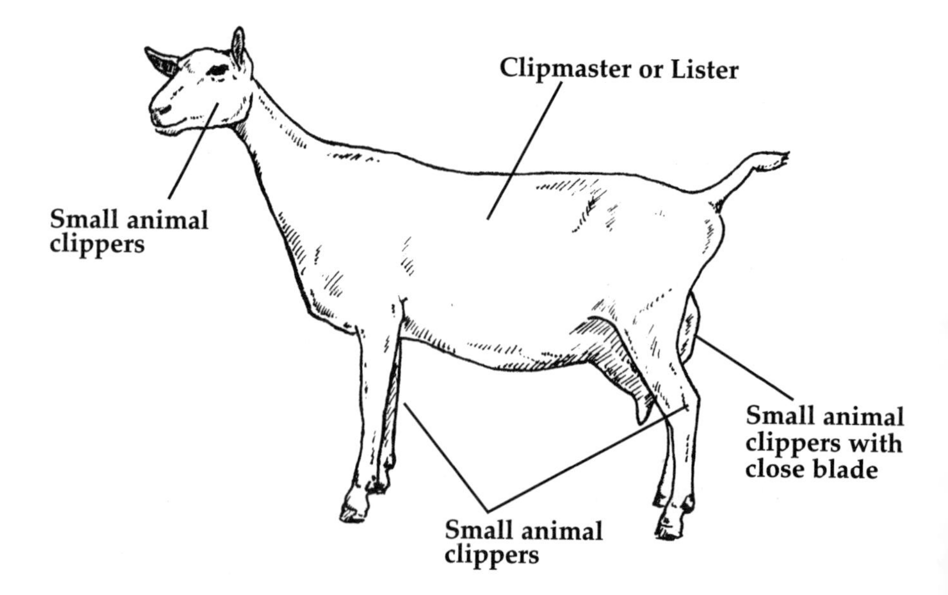

of the doe. If you use only one clipper, use the standard blade provided with it.

Before clipping the doe, give her a good brushing or wash her if the weather permits. This removes dirt from the hair and will give your blades longer life.

Also remember to lubricate the blades regularly. Use a commercial blade lubricant or a mixture of one part motor oil to one part kerosene.

For an even, uniform clipping job, always clip against the lay of the hair. Your objective is to get a close clip of uniform length over the entire body of the doe.

Clip the hair along the edge of the hooves and under the dewclaws, clip the hair out of the ears, clip the tail, face, and inside of the legs, and on the chest floor.

When clipping the head, be sure not to clip off the eye lashes. They protect the eye from foreign matter, and if you remove them, the doe may develop an eye irritation. It is not a bad idea to put eye drops in the eyes after clipping to avoid any irritation. Dog groomers rub vaseline around a dog's eyes before clipping to keep the hair from falling in the eye.

There are three ways to clip the tail: tufted (a tuft is left at the tail end), clipped (the tail is clipped completely), or square (the hair is "squared off" at the tail's end).

What part of the goat you choose to clip first is up to you. I start with the legs and tail, do the body and neck, and then finish up with the udder and head. If you use a close blade on the udder, do this at the show, and be sure that there is at least twelve hours of milk in the udder. If the udder is not fairly tight, you may knick the skin. If your doe has a beard, it needs to come off.

Clipping the Buck

Bucks are clipped the same as does except that you do not remove the beard, especially if it is long and full. A Nubian buck with a scroungy beard, however, may look better if you take it off.

Trimming Hooves

Perhaps the most ignored part of dairy goat management is keeping the hooves properly trimmed and shaped. You cannot correct grossly neglected hooves in one trimming prior to a show.

Remember that your primary objective when trimming hooves is to keep the depth of the toe and the heel the same. Many people trim the heel too short and leave the toe long, causing the animal to rock back on its feet. This may cause the pasterns to look weak or broken.

You can do some corrective hoof trimming to adjust for weaknesses in the feet and legs. Remember, however, that such corrective procedures will not cure structural problems nor completely correct a weakness.

Animals who tend to turn in at the hocks should have the outside half of the hoof trimmed shorter than the inside half. This will cause the animal's legs to bow out somewhat and increase the width between the hocks. The difference in depth of the two halves of the hoof should not exceed 1/2 inch or there will be too much stress put on the bones and joints.

Goats whose front feet toe out should also have the outside half of the foot trimmed shorter than the inside half. Again, do not exceed more than 1/2 inch in difference of depth.

Final Grooming Procedures

Perform final grooming procedures on show day. These include cleaning the coat, washing the hooves, and applying a coat conditioner. If the weather permits, wash the goat with livestock shampoo. This will remove dirt and stains as well as give the hair a soft texture. If you can not wash the goat, you can use a damp cloth with a little soap to clean dirty areas. Pay particular attention to knees and hocks.

If the animal has dandruff or dry skin, wash the goat well in advance of the show to remove the flakes of skin. I often use baby oil to condition the skin, but you will need to rewash the goat before the show because the oil attracts dirt. Many livestock shampoos contain skin conditioners, so read the labels before making your choice.

A word of advice. If the animal has been treated with a wound dressing and the stains are still on the hair, consider leaving the goat at home. Exhibitors may be wary of the animal's condition, and it is just much easier to take her out to show when there is no question about her condition.

I usually apply a show spray to the coat right before taking the animal into the ring. This will usually give the coat a glossy appearance, and most of these sprays contain fly repellent. (After showing animals on a hot summer day, you will learn that fly repellents are a must!) White areas stained with urine or manure can be touched up with talcum powder or whiteners intended for horses or dogs.

If you are entering a showmanship contest, your animal must be spotless. This requires washing and scrubbing the hooves to remove manure. Some exhibitors polish the hooves to make them shine. If you do this, be sure to use the color that matches the feet. There is nothing more unsightly than a Saanen or Toggenburg whose feet have been polished black.

I prefer to use a small amount of vaseline on the hooves. I remove any excess with a dry cloth. This will give the hooves a fair amount of shine. If you do not remove the excess vaseline, dirt will stick to the hooves, so be careful.

Use a damp cloth or packaged towelettes to wipe ears, nose, and tail. While all of this is not strictly necessary if you are not showing in a showmanship contest, it is truly pleasing to see an animal enter the ring that is spotlessly clean. And while these grooming procedures will not change the conformation on a goat, the little extra may catch the judge's eye and make just that subtle difference.

Showing Dairy Goats

This article appeared originally in the June, 1983, United Caprine News. *We wish to thank Jeff Klein, publisher, for allowing us to reprint an edited version of it here.*

Showing Dairy Goats
by Jeff Klein

The American Dairy Goat Association Showmanship Scorecard reflects ring procedures that have been established to add uniformity in the show ring and to aid in keeping a show moving smoothly. Imagine the chaos in the show ring if each exhibitor used a different procedure when exhibiting his animal!

Before entering the show ring for the first time, it is wise to familiarize yourself with both the general and showmanship scorecards. However, expertise will come only through practice. Watch the more experienced handlers in the show ring. You can often pick up a few pointers by watching them.

Since showmanship is the presentation of an animal in the ring to make it look its best, studying the ADGA scorecards and Evaluation of Defects will help you evaluate the animals you are showing and guide you in knowing how to show them to best advantage.

While being a good showman is important, even the best showing procedures will not influence the final placement of the animal to any great extent. It is impossible to hide faults on an animal while it is moving around the show ring, and for this reason most experienced judges do most of their evaluation as the animals are walking, not while they are standing still.

Keep in mind that your goat must be acquainted with your handling techniques. Young kids usually learn very fast—just a few minutes of work a day will usually do it. Older animals that have never been worked with will take a little more work.

The only equipment you will need to show your animal is a collar. Do not use a leash or lead. Small link chain collars are the best but leather will work. If you choose to use a leather collar, use a thin one. The idea is to present an animal that looks sharp and clean. Thick collars detract from the animal's neck (which should be long and lean). Size the collar to the goat. Thick plastic link chains are fine for mature does and bucks, but look over-large and detract from kids. The showmanship scorecard specifically calls for a "collar of small thin chain, properly fitted."

It is customary for dairy goat exhibitors to wear white, and some shows may require it. White pants and shirt are the proper dress. Sandals, shorts and halter tops may be appealing on the street, but they should not be worn in the show ring. Remember that the goat should be the focus of attention, not you!

Showmanship Procedures

The most important thing to keep in mind when showing a dairy goat is that the animal must always stand between the exhibitor and the judge. The reason for this is simple: it will always give the judge an unobstructed view of the goat.

To accomplish this, you will have to switch from one side of your animal to the other as the judge moves. Always switch sides by moving around the front of the animal so that you can maintain complete control. If you switch sides by crossing behind the goat, your arm will be stretched over its back, and if the animal decides to run, you will not be able to hold it back.

Keeping the animal between you and the judge at all times may be confusing at first, but with a little practice you will learn how to do it smoothly. Diagram A, below, shows the proper procedure to use when the animals are being moved around the ring and the judge is standing at the outside of the circle. Notice that the exhibitor is always on the opposite side of the goat from where the judge is standing. The handler must switch from side to side of the goat two

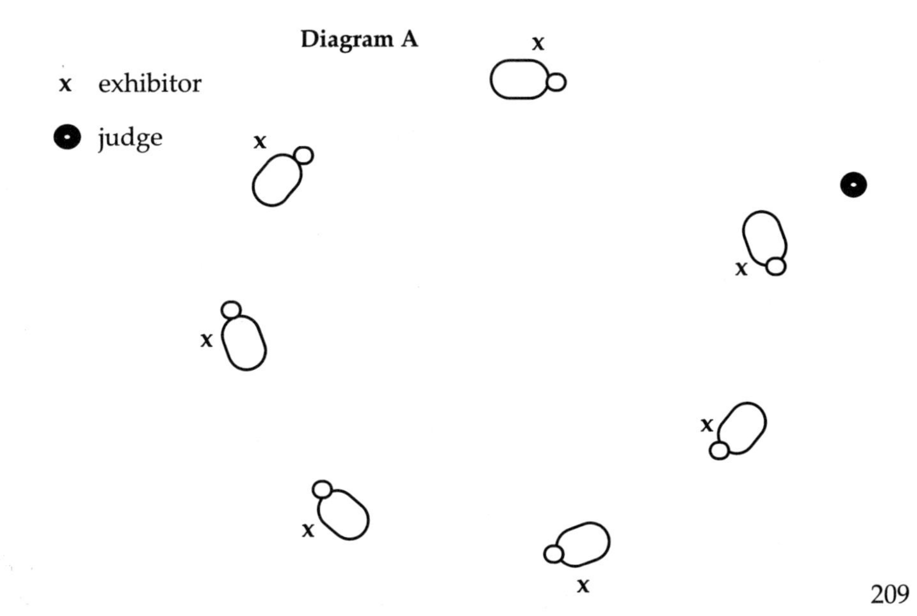

times—indicated by the two arrows. Again, notice that switching sides is always done around the front of the animal.

If the judge is standing in the center of the circle, the exhibitor will never have to switch from one side of the goat to the other since the handler will always be on the correct side of the goat.

When animals are lined up side by side, the handler should always leave enough room between goats to allow ease of movement. Sometimes this is difficult in a large class showing in a small ring, but part of showmanship is being courteous to other handlers.

As the judge moves up or down the line (Diagram B), you will have to move from one side of the goat to the other to stay in the proper position. Never stand at the rear of the animal, even if the judge is at the front of the line. Simply stand to one side of the goat to allow the judge a full view of the animal's forequarters.

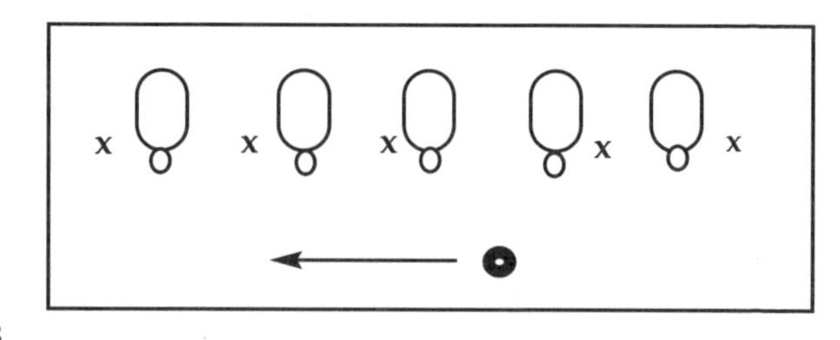

Diagram B

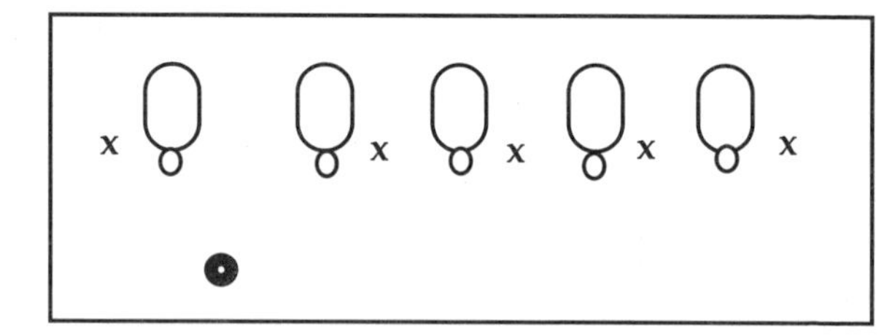

When asked to line up goats head to tail, be certain that you leave ample room between animals. There is nothing more frustrating than to set up your animal, only to have the goat behind nibble its tail. Your goat will not stay set up correctly for long!

If the judge moves from the front to the rear of the line, depending upon where the judge stands, you may find yourself on the wrong side of the goat. Switch sides smoothly by moving in front of the goat. If the judge moves from one side of a head-to-tail line to the other (Diagram C), all exhibitors must switch sides.

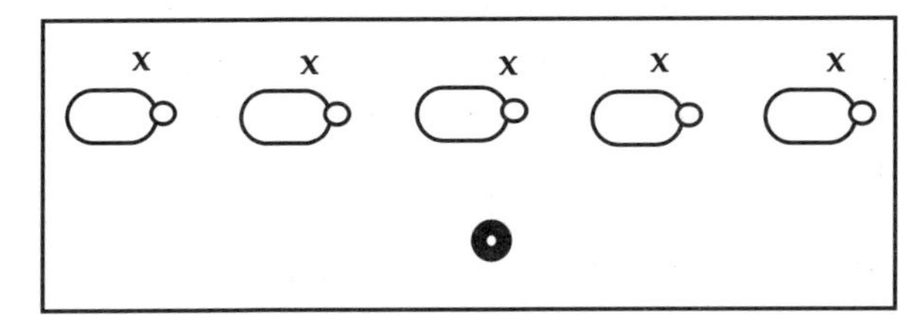

Diagram C

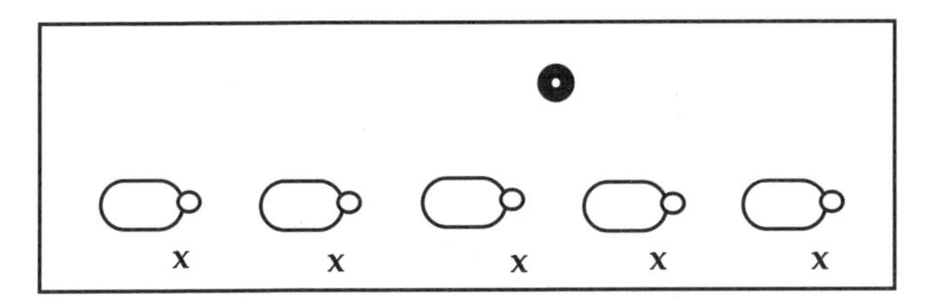

If the judge asks you to move your goat up or down a head-to-tail line (Diagram D), always move the animal in front of the line so that the judge can see it clearly as it moves. Again, be sure to keep the goat between the judge and you. Other handlers should make room for you in your new position.

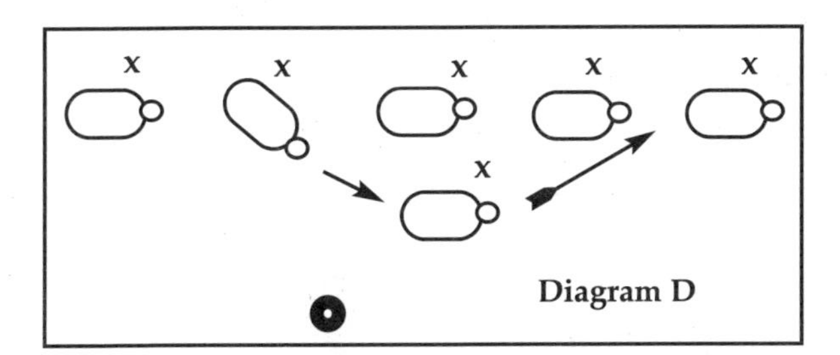

Diagram D

One of the most difficult showmanship procedures to learn is moving an animal to a different position in a side-by-side line. It is even more difficult to describe in words so you will learn best by watching it performed by a skillful showman.

When moving into a different position in a side-by-side line, always lead the goat out to the front and walk up or down in front of the line of goats to your new position. This gives the judge a chance to see the animal in motion, and it also allows other exhibitors to see what is happening. If you were to lead the animal behind the line of goats, that judge could not see you, and exhibitors might not notice that they need to make room for you in your new position.

Moving up or down in front of the line, lead the animal to the position requested by the judge and walk through the line to the back. Then make a U-turn and finally bring your goat back into position. While doing this, you must remember to stay on the proper side of the goat (away from the judge). See Diagram E.

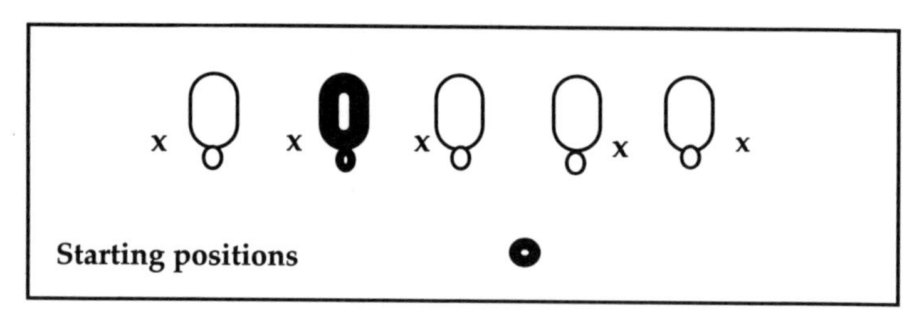

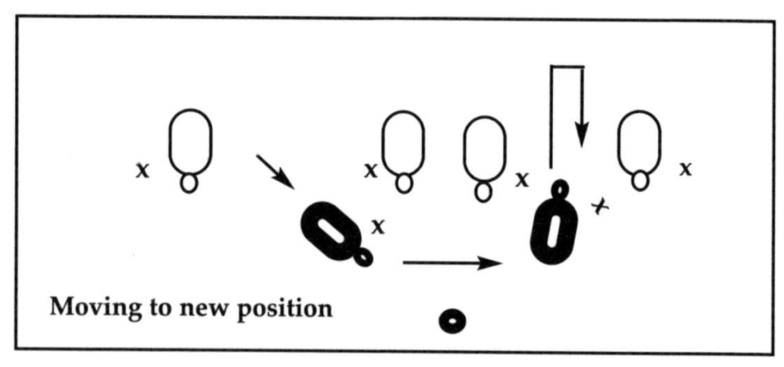

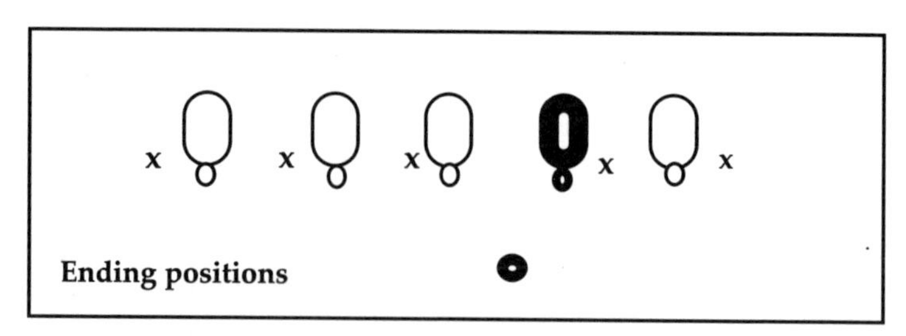

Diagram E

Presenting the Dairy Goat

Knowing proper show ring procedures is only half of the showmanship picture. You must also be able to present your animal correctly. This means you must set up your goat to best advantage and in a way that it resembles the ideal dairy goat as much as possible.

Most dairy goat exhibitors tend to spread their goats' legs too far, giving the animal a rocking-horse stance. This will often cause the topline to sag.

The goat's head should be stretched up and forward so that the shoulders and withers will blend smoothly and appear sharp. However, if the goat's head is held too high, the topline will sag, and the shoulder area may look loose.

Avoid placing the goat in an exaggerated position. Set legs squarely beneath the body. You should be able to draw a line from the pin bone to the hock, and this line should be perpendicular to the ground.

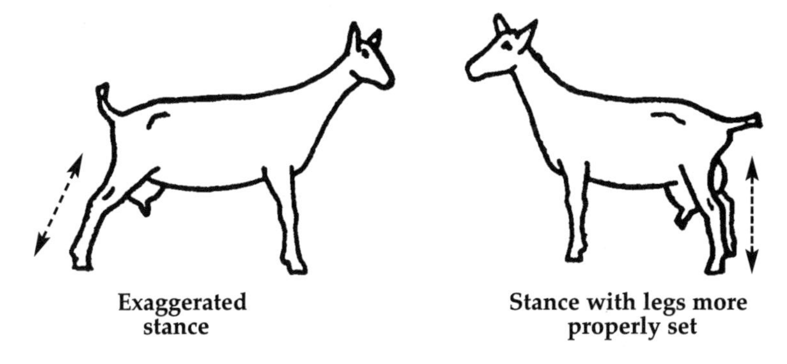

Exaggerated stance **Stance with legs more properly set**

Handlers also need to make sure they do not set the rear legs too far apart. Doing this with milking does causes the udder to look as if it lacks capacity since there will be space between the udder and the legs. While there should be plenty of room between the rear legs, it does not help to overdo it. If you do, the judge will probably look at the animal's legs more closely as it moves around the ring because he may think you are compensating for a weakness.

If your animal's legs turn in at the hocks, you can move the hocks out while she is set up. However, as you move around the ring, there is nothing you can do to hide close hocks, and an experienced judge will see this fault.

Weakness in the chine or loin area can be corrected by scratching the goat's belly. This will cause her to raise her topline and make it look straighter. Sloping rumps can be corrected by scratching the animal in the hip bone area. Remember, however, that most judging will be done on the move, where you really can not do much about a weak chine, loin, or other structural defect.

All of your movements, including pushing, prodding, scratching and setting up, should be done calmly and quietly. Do not whistle at the doe or make noises to keep her alert. If your goat's attention tends to wander, simply give her a sharp jerk with the collar. Also, do not snap your fingers or throw sawdust to get your goat's attention. That will work fine for taking pictures but is inappropriate in the show ring.

Remember that the most successful showman is the one who is not noticed in the show ring. You want the judge to focus on the animal, not the exhibitor. Your goal is to make your goat the best it can be.

Showing Boer Goats

This article was originally published in the February, 1998, issue of the Goat Rancher *magazine. We wish to thank Terry Hankins, editor and publisher, for allowing us to reprint an edited version of it here.*

Showing Boer Goats
by Cathie Keblinger

The Right Stuff

Lacey Gannon, a high school sophomore from San Angelo, Texas, is demonstrating how to prepare a Boer goat for showing. She has placed a blocking table in the middle of the ring. On it she has a plastic storage box filled with the items she'll need on show day.

First she pulls out a folder filled with the goats' health papers and weight books.

Next come the leads. She uses a 12-inch leather loop with a snap on it. It attaches easily to the neck chain on the goat. The leather is comfortable on her hands, especially if the goat pulls on it. Lacey used to use a halter on her goats but now prefers neck chains after one goat wearing a halter nearly died when tied to a fence.

She also has a few three foot long rope leads with snaps, some extra 2-way snaps, old towels, a pair of little scissors, and some syringes.

Lacey makes sure she has a drench gun and packets of tasteless goat electrolytes, and a muzzle to keep her goats from eating shavings or someone else's leftover food.

To prepare her goats before the show, she has soap, waterless shampoo, and a brush, a little hand sprayer that can be used with a jug of water, Alberto VO5 to condition the coat after washing, and Showsheen to spray on and wipe off right before showing. Her Hurricane blower will blow dry her wet goat, and sometimes if her goat is especially clean, she can use it (without washing) to blow the dust out of his hair.

Since they usually shear their meat goats for show, Lacey has a goat-shaped blanket for cold weather. A jersey bodysuit under the blanket will protect the goat even more. For especially cold conditions, she has a small propane bottle with a little heater attachment.

While all of this equipment is not necessary for exhibiting at one or two shows close to home, if you are going to exhibit at a number of shows, you should have a show tack box with everything you'll need both for showing and for emergencies.

Preparing the Goat for the Show

Lacey uses Lister shearing equipment to groom her goat for show. This shearer is a good one since children can use it safely. She shears the goat all over if short hair is required, and trims the hair on his tail and above his hooves just to the top of the hooves.

She pays particular attention to trimming her goats' hooves, as keeping hooves trimmed right is of paramount importance for the goat to stand correctly. She thinks it's almost an art form.

Getting goats to drink just the right amount of water can be a problem at a show. Since meat goats are shown in weight categories, if a goat drinks too much water, he may have to show in the next weight class, and "Nobody wants to be the lightest goat in the heavy-weight class—I'll promise you that!" Lacey says.

She uses a drench gun to give the goat enough fluids without letting him drink so much that his stomach gets big or he goes over into the next weight category. She trains her goats to drink water with electrolytes at home. She also trains them to drink when she puts the bucket in the pen. To monitor their water intake, she'll put a bucket in the pen two times a day for thirty minutes. If the goat is close to the weight limit, she'll drench him with water and electrolytes a little at a time, which will keep his fluids up, but she'll offer no free-choice water. Lacey says it's important to understand these strategies as the rules allow only a four-pound fluctuation from the weight on the weight card.

Success In the Show Ring According to Lacey

To be successful, you need four things: You must have a good goat, you have to take good care of him, and you must know how to show him. The fourth part is luck. According to D'Wayne Gannon, Lacey's dad, "If the goat and the judge come together like two magnets, you've got it."

Showmanship, however, is tremendously important. You're in the ring to convince the judge that you have the best goat. Lots of shows are won by showmanship, according to the Gannon family.

The first impression the judge gets of the goat as it enters the ring is especially important. Sometimes the judge's best look at the goat is when it enters the ring. Lacey tries to pull up a little on the lead as she enters the ring so that the wether's muscles will ripple just a little. That may catch the judge's eye.

If you have a good goat with a lot of balance, the judge might pull you out of line as soon as you enter the ring and put you in the top group. Many times, according to Lacey, the judge will say in his summary that the champion goat caught his eye when the goat first walked into the ring.

The cardinal rule for showing goats is to never stand between the goat and the judge. You must also always watch the judge and stand on the side of the goat away from the judge. If the judge asks you to lead the goat toward him, be sure to walk the goat directly toward the judge, staying on the side away from the judge. Turn the goat when the judge tells you to lead him away, and again make sure the goat is nearest to the judge. Walk directly away from the judge so he has a straight view of the hindquarters, then turn, again with the goat between you and the judge, and set the goat up.

To set up her goats to look their best, Lacey spends time studying her goats in their pen at home. She wants to see how they stand naturally. Then she'll practice setting up her goats that way and try to set their feet that way in the show ring.

"Some goats look better with their hind feet a little farther back, but some of them look good with their hind feet almost straight under them. You just have to know your goat and adjust to his style."

The front legs should be straight under the goat and out at the corners of his body. You can lean over the goat to lift his feet into place, or if the goat is small enough, you can lift his front end off the ground and let his legs fall back down. Either way works well. To make your goat move his hind feet back a little you can push on his hips. Sometimes pushing on his chest has the same results. Training and practice at home will allow you to perfect your showmanship technique.

Whether the goat is standing in place or walking, you must always keep his head up.

Knowing your goat's strengths and weaknesses is all important. Always place him in a stance to emphasize the good points and minimize his faults. If your goat is a little too fat and has fat behind his forelegs, you can help hide it by moving his forelegs slightly forward. You can make a narrow chest look a bit wider by moving the foreleg on the judge's side back a few inches.

Once you have set up your goat, you should move back to stand even with the goat's shoulder, slightly away from the goat's body. The judge will then have a clear view of the goat's profile, including the head, neck, and chest, and can see the goat all the way around—even on the side where the exhibitor is standing.

According to Lacey, "One of the most important things is not to let the pressure get to you. If you do, it will cause you to make mental mistakes. I know because I've done it many times."

Since small mental lapses can cost you the championship, to be a good showman you need to pay constant attention to the details that make up good showmanship. And Lacey has certainly done that as she's won showmanship countless times and has an impressive number of championship wins for her goats also.

Hints

"For longer clipper and blade life, always put your clippers away clean. Brush away accumulated hair and spray with clipper lube before you store them."

"Don't trim your goat's hooves at a show. Trimming should be done a few days before the show to give them a chance to get used to how the foot meets the ground. It's always possible to trim too close, and there's nothing worse than having your best doe limp around the ring just because you've trimmed her hooves too close."

"If your doe has trouble walking around a full udder, it may help to put some udder cream ointment on the inside of the rear legs and on the sides of the udder. This can help the udder slide by the leg more smoothly."

"While we do like winning traditional trophies, we've found that winning can be memorable every day of the year if we milk in a 9-quart pail we won for best in show, or if we wear a belt buckle our buck won for grand champion. And we really appreciate the hoof trimmers we've won. Most traditional trophies gather dust, but a bucket or hoof trimmer never will."

"Idle Hands Are the Devil's Workshop"

This article originally appeared in the Dairy Goat Journal *in 1996, and we reprint it here with our most sincere thanks to Dave Thompson, publisher.*

Idle Hands Are the Devil's Workshop

by Joan Vandergriff

"Take a book along wherever you go," my mother would remind me, "in case the dentist (doctor, lawyer, Indian chief) keeps you waiting." She taught me well. To this day, when I forget to do this, I'm always sorry. No waiting room literary fare can substitute for that really good book left lying on my bedside table.

Rarely, though, even if I've forgotten to take along reading material, do I find myself bored. People watching, quiet musing, talking to strangers can fill my idle time.

My husband seems to be less able to keep himself amused. He, too, was well-taught by his mom (and reminded by his devoted wife) to bring along a book to read to fill those yawning hours. But for him, books can hold his attention only so long. For him, people watching is too passive, musing too internal, and talking to strangers too close to his sister-in-law's behavior. (She can cite a person's whole personal history after a two-minute conversation in the grocery line.) He needs action to keep him occupied.

At a past national show, you may remember a severely bored husband who kept himself and others occupied using a fishing pole with a dollar bill attached. As he cast the bill out in the aisles, unsuspecting people would bend to pick it up. At the moment before hand met bill, he retrieved the dollar on the string. Big jump, big laugh. He said it worked best with kids.

His creative answers to boredom were not always so frivolous, however. I remember how boredom once led him to his first foray into improving a goat's appearance through cosmetics.

It was the 1984 National Show in Pomona. We had brought four goats, all two year olds: two black beauties, a honey-colored lovely, and a mousy beige wallflower. It took little work to keep the goats well-cared for, and our display tables were manned by others most of the time, our books were at the motel, and the breed showing held little interest. The perfect combination for either inspiration or devilment.

How he got the Lady Clairol—how he even got the idea—I don't know. But the afternoon before we showed our goats, he said to me,

"Take a look at Mousy. Whattya think?"

I wandered down our aisle, located our pens, and couldn't for a moment place Mousy. I counted two black, one honey-colored, and one richly colored dark beige eye-stopping goat. Mousy was the same goat, almost the same color, but her coat was ever-so-slightly darker. The color looked as though it went down to the skin. She stood out from her surroundings. She caught your eye. And most important, she still matched the description on her papers.

As I examined her more closely, I realized that her very best feature, her long straight back was prominent. "How'd he do that?" I thought. Coming up behind me, my husband obviously had overheard my silent musings. "How do you like her dorsal stripe? I used a little horse paint to darken it down a little. Really makes you see her long, straight back."

I looked to her hooves, which were still the same old color.

"I didn't paint her hooves 'cuz I didn't want to draw attention to her weak pasterns," he said, obviously pleased with himself, and then he drew me over to one of the black does whose hooves gleamed.

"Footsy's hooves are her best feature. Thought I'd try a little paint on them. Really draws your eye to 'em, doesn't it?"

"Cheating," say the purists. "Off with their heads!" But I don't think so.

As a groomer and handler of show animals, my task is to show each animal to best advantage, and that begins way before we ever hit the show ring.

Having been around show rings of dogs, horses, cattle, sheep, pigs, and other assorted living creatures, one thing is evident—when you reach a certain level of competition, you can't just wash your animal, groom it, put on a fancy collar, waltz into the show ring, and expect that the animal will necessarily win on its own inherent merits.

Pretty cynical? I don't think so. It's just common sense. Even the most beautifully put together goat may not "shine" without a little help from her groomer. And here's where ethics really come in to play. As you work to make your animal all she can be, she must still be the goat you began with, the goat that matches her papers.

What you're trying for is that little edge that will make your goat stand out—positively—as she walks in front of the judge.

To do this, you need to learn a few tricks. And you need to learn how to use them without entering that area that we consider unethical.

The Clairol trick can work with just about any color goat. And you don't need Clairol, pick whatever brand you're loyal to. On freshly clipped black goats whose white skin shines through, try the darkest black you can find. On dingy white goats, those that look as though they have a "nicotine" yellow tinge to them, try a blond which has a little "blue" in it.

Above all, just as you would on your own hair and skin, be sure to try a "patch test" before you apply any chemical. You never know which animal (or human) might be sensitive to a product or chemical.

You now know how we used the horse paint to bring out an animal's dorsal stripe, and it can be used to cover skin scars that mar an otherwise smooth coat or other unsightly marking. You must not use it to cover defects such as white spots on Toggs or black spots on Saanens. Ethics, again. What you're trying to do is to make your lovely goat look better, not trick someone into overlooking a serious defect.

The dividing line isn't always clear. Take the breeder whose dark goat had a white belt that cut the goat in half, making her look short coupled. Spraying the belt to match the rest of the coat would make the goat not match her papers, but lightening it up? Best advice is if you're worried that what you're doing isn't right, then don't do it.

As we saw above, hoof paint can brighten up those dull-looking feet, drawing attention to them. Decide whether the hooves are a positive quality of your goat before you do it. At the very least, you can use a shoe polish to shine those hooves, so they look clean and neat.

Clipping to enhance one's animal is truly an art. If you've been around cattle or sheep people, you've probably seen them spend long hours with clippers and brush in hand. What they're doing besides getting rid of extra hair is to literally "sculpt" a shape to their animal that emphasizes positive qualities and hides negative qualities. Thus you'll see that an animal with less than full haunches will have its hair fluffed out and trimmed and sprayed in place to give the haunch a fuller appearance. Such grooming is a given when showing these animals.

In goats, it's harder to do this kind of grooming. First, a goat's hair isn't like a sheep's, making it difficult to trim, fluff, and stand up. Second, for the most part we want to make our goats look less round and full rather than more, so we don't need the same sort of grooming techniques as sheep people do.

In grooming a goat to emphasize its qualities (or de-emphasize them, as the case may be), we need to learn how to leave hair on where fullness and width are needed (such as through the legs as seen from the rear) and how to trim hair short (such as along the throat and neck) to give a leaner, cleaner look. Finally, we need to be able to blend longer and shorter areas so that the animal looks smooth, not a hodge-podge of short and long hair.

The key to learning how to clip your goats to maximize their potential is to really know each one's strong and weak points. Whether you do this by just looking at them critically yourself, or you let a professional help you (a judge, appraiser, or breeder with a good eye), you can't groom your animal well until you know just what needs enhancing and what doesn't.

The most basic level of such grooming is learning just how long a time each goat needs to grow a coat that will make it look its best. Most exhibitors will clip their goats a few days before the show, feeling happy if they just get the clipping done in time. However, to really do a good grooming job, you need to know if your animal looks best freshly trimmed (the night before the show, for example) or does it need four days or more to allow the color to fill in and block the light skin from shining through.

My Toggs always looked best if they were clipped right before they were shown, and my light-colored LaManchas were the same. My black goats always needed far more time to "grow back in" in order to cover unsightly clipper streaks, scars, coat unevenness, and skin. Showing goats with such blemishes takes away from the animal's overall appearance, and draws the judge's eye away from the total package and first impression that often makes a difference in top competition with lots of nice animals in contention.

Udder trimming is the one area, I think, that separates the novices from grooming pros.

When I first began showing, I learned to trim udders in the usual fashion. Remove any long hair around the udder with a regular clipper, then using the closest trimming blade possible, "shave" the entire surface of the udder. Period. I'd do this the day before the show, and I'd always wonder why my goats' udders always looked a little overgrown and scruffy when compared to those of more experienced exhibitors.

Over my showing career, I learned that each goat's udder demands individual attention. Depending on the shape, height, fullness, width, and length (and probably other characteristics), I would trim to emphasize its best points and minimize its faults.

When clipping goats before a show, I go over the udder with a regular blade. I save the final trimming until the morning of the show when the animal has a fairly full udder and I can see just where I need to trim and where not to. In general, if a doe has narrow or low rear udder attachment, it's best, especially on dark goats, not to "skin" the udder down excessively. Excessively light udder skin against the dark hair on the legs draws the eye to the udder, allowing the viewer to see its narrow width and saggy rear udder attachment. Unless your animal from the rear has good udder width and height, play it safe by trimming the udder normally with a close blade, removing any long hair and cleaning up the udder surface. Do the same for the front attachment.

When I first began showing, on does with genuinely pocketed fore udders, we would leave udder hair in the front and on the belly untrimmed, hoping to "fool the judge." Did it work? Probably not. For those animals with less than perfect fore udders, use the same strategy as on the rear udder. Clean it up but don't do anything that will overemphasize an imperfection.

One of the most difficult grooming tasks to master, in my opinion, is what I call "the National Show udder trim," where the goal is to remove every single errant hair, leaving the udder as smooth as the proverbial baby's bottom. Part of this task is removing hair around the rear attachment, showing off the natural arch.

The problem here is that the area to be clipped is round and the clipper blade straight. A slight miss-trim and you have a gouge or "v" which immediately becomes the focal point whenever you look at the rear of the goat. This trimming fault becomes especially obvious on dark-haired animals.

There are two ways to overcome such trimming errors: one psychological and the other cosmetic. The easiest thing to do is to convince yourself that only you, the one who did the actual trimming, can see an imperfection; that no one, neither judge nor viewer on the sidelines, will ever notice the slight jaggedness to the trimming job. This works well if you win the class or win first udder. If you don't, you tend to blame the lousy trimming job, no matter what the actual reasons are.

The cosmetic trick is a bit more difficult both to learn and to do. One word of warning: this works best on black or dark-haired goats. Trim your goat's udder the morning of the show as you would normally, using a close blade or whatever trimming device you use. As you remove the hair around the rear attachment, you don't need to be as picky since you are going to fill in any miss-trims with spray paint. Make sure that you remove more hair rather than less. If there are any gouges or "v's," don't worry.

To clean up the trim on the rear udder, you need two things: horse paint that matches your goat's hair color and a template. To make the template, use scissors and light cardboard to cut a curve which approximates the shape you want to see on your goat. A further caution: make sure the curve you cut follows the natural shape of the udder. All you are trying to do is to fill in a faulty udder trim and make the udder stand out, not change the lines of the udder. Next, hold the template against the udder and spray a fine mist of paint over the miss-trimmed area. Be sure you don't overdo it. Too much paint can make a sticky mess.

For the fore udder, you can use the same paint without a template. Simply use a very fine spray to smooth the margin area between where the udder trimming ends and the stomach hair begins. Use your hand to smooth out the paint as needed.

Finally, after the paint has dried, apply a skin softening cream or conditioner to the udder surface, being careful not to smudge the paint.

Skin and coat conditioners can help with overall appearance also. There's nothing more appealing than seeing a goat that looks "squeaky clean" and shiny. Also, judges like their choice of top goats to look the best to those who are watching, and a poorly groomed, less than perfectly clean animal may not do that.

When judges put their hands on animals, they want coats to feel smooth and clean. Some judges don't like the feel of certain products that make the hair slick, and no one likes to feel a coat that is sticky or stiff, so make sure the products you choose not only make your animal look good but feel good to hands-on judging as well.

Also, be careful about using "glitsy" products such as glitter, iridescent spray, or perfume. Those are fine for dramatic presentations (perhaps a sale or costume class), but in the show ring they usually detract from the overall appearance of your animal.

The final "trick" has nothing to do with cosmetics or grooming and everything to do with proper feeding and husbandry. No animal can look its best if it hasn't had the best care prior to coming to a show.

Some people feed their show animals differently than their non-show goats. That's up to you. However, the bottom line is that if you haven't prepared your goats with lots of good feed, water, and minerals long before the show, you can't make up for this lack during a few days of special care at a show.

And I can't tell you what feed will make the difference for your animals. Each part of the country is different. In our area, we couldn't feed all alfalfa because it caused mineral imbalances that we couldn't overcome. Therefore, we had to develop a feeding program that allowed our animals to prosper and bloom without encountering other problems, including obesity. We did this through a mixture of different hays, grains, and minerals.

There are things you can do at the show, though, to ensure that your animals, under the stressful conditions of the show, don't lose the bloom you've worked so hard for.

Bring enough hay and grain from home so that you don't have to change their diet prior to showing. Make sure that they have fresh, clean hay and water in front of them at all times. If you have goats that are picky about the taste of water (to be honest, I've never had any of these—if they get thirsty enough they seem to be willing to drink anything!), use an additive that they find tasty. But try the additive out on them at home since a show is not a place for surprises.

Some people will feed kids a large bottle of water just before they enter the show ring in an attempt to give them more substance and width. I've never found this helpful as it really looks weird to see this rounded water bottle of a kid waddling around the ring, making sloshing sounds (I kid you not!) as it moves. With its skin as tight as a drum, the kid can't be comfortable, and the judge will find it difficult to feel its openness and skin texture.

With milkers, feed sufficient water and grain to allow them to make milk under the stressful conditions of a show, but also remember that drastic changes in feeding may lead to stomach upsets and the opposite effect of what you intend.

We all know the adage about silk purses and pigs' ears. Our goal here, then, is not to turn a poor specimen into a show winner (an unrealistic goal), but only to make the most of the positive qualities in our animals and minimize the less positive. Surely we can put our free time at shows into more useful pursuits than my husband's fishing expedition. He really should stick to developing a line of goat cosmetics!

An aside: Does my husband have any other "tricks"? Well, he *could* tell you how to produce a more prominent medial suspensory ligament and how, in ten minutes, to increase udder attachment, but don't ask him to share his secret. These two tricks are unethical.

"On the Road Again"

This article originally appeared in the Dairy Goat Journal *in 1996, and we reprint it here with our most sincere thanks to Dave Thompson, publisher.*

On the Road Again

by Joan Vandergriff

SCENE 1: A lonely rest area on the interstate somewhere in southern Oregon. Ice pellets mix with sleet slivers to form a transparent blanket engulfing the entire little blue car. No one can see in. We can't see out. It's hard to find anything good about the situation until one looks outside the car and sees the trees crusted with ice jewels. I'm not sure I have ever seen anything more beautiful or more magical than this winter ice storm. And yet, it's still hard to find anything good about the situation.

We had to stop. The defroster couldn't keep up with the ice. Our tires couldn't find dry road. Not so bad you say? Well, it wouldn't have been with just the two of us. Unfortunately, we are three: me, my husband, and our newest Toggenburg doe, whom we've just picked up during one of the worst ice storms in Oregon's history.

Stranded for the night in the rest area, we begin with my husband and I in our front bucket seats. We're afraid to run the engine much; afraid we'll run out of gas; afraid we'll asphyxiate ourselves. Afraid we'll freeze to death, we climb into the back of our hatchback and cozy up with the goat. She's a warm natural furnace; she's chewing her cud; she's happy. We doze on and off, waking to the repeated belch of cud. In the morning everything inside the car is wet from the humidity of our natural heating system.

SCENE 2: A busy airport somewhere in the midwest. I'm on my way to the ADGA National Convention with my Spotlight Sale doe. Her crate won't fit on the baggage cart. She is delivered *sans* crate by elevator and must be led the length of the airport to hook up with her crate. It's -10°F outside. We're walking through the passenger lounge, a very nice man follows behind with a broom and a dust pan. He uses both frequently.

SCENE 3: The Nevada County Fairgrounds after a three-day show. I'm there by myself, and I'm slow to load up. Every other exhibitor has now left. I've got six kids and as many milkers to load in a two-horse trailer. The kids go up top, not a problem. The milkers go down below. I get three in—three more to go. The next one balks at the door, two push out, one goes in. I round up one escapee. She goes in but one more jumps out. One goat comes over and pulls my hair. At some point I sit on the ground and cry. I eventually round up the strays, push them in, and double lock the door. I do not get the kindness award on this trip.

Even if you have only a few goats, you'll probably have to transport one or more at some point, perhaps to the vet or a breeder. Even the shortest trip can pose problems for you and hazards for your goats. And if you're shipping animals across the country or around the world or trailering them to another county or another country, transporting goats can be a trial. What's worse is that there are a few factors, like weather, that you have little control over, so it's wise to have a basic strategy that will head off some potential problems. The following advice works for me whether I'm shipping one kid by air or transporting a herd by trailer.

Keep the sick ones at home.

When sending a kid to its new owner, remember that a sick kid will only get sicker in a crate on an airplane or in a car or trailer. Don't rely on your vet's check and health papers to certify good health. A lot can happen between when you have the animal at the vet and when you ship it. The night before you plan to transport the animal, check it over carefully. Make sure its eyes are bright, not matted, that there's no sign of a cold. Its coat should lie flat, not stand up, indicating a chill or fever. Be sure there are no signs of diarrhea. Do the same in the morning. If the goat has an obvious problem or you suspect it may be coming down with something, postpone shipping it. A goat will never experience a "miracle cure" spending a day in a trailer or in a shipping crate in the baggage compartment of a plane. Nor will giving it a prophylactic shot of some antibiotic ensure a healthy result. More important, the antibiotic may mask symptoms of a serious health problem.

Do the same health checks for animals you take to shows or to a breeder. A seemingly minor problem can turn serious quickly under stress, and traveling for animals is a stressful situation. Also, problems that wouldn't spread to other animals at home can become an epidemic when animals are crowded in an unfamiliar environment. Remember how angry you were when your goats caught the skin problem from goats penned next to yours. Even your best friends won't like you if their goats catch something from yours.

The weatherman is your friend.

As we found out being stranded in the Oregon ice storm, bad weather can make traveling with animals a nightmare, sometimes even dangerous. When you ship animals by air, the airline may determine when the weather is too severe to risk shipping live animals. Often during the summer, temperatures in the West, Southwest, and South, especially, must be below a certain level before shipping agents will accept animals. When temperatures come close to these levels, even if airlines allow shipping, animals can suffer. Consider getting animals shipped to these warm areas prior to the hottest months, try shipping on the earliest flight in the morning, or use alternate transportation.

The cold poses similar problems. Remember that a single animal in a crate has nothing to snuggle up with to help maintain its warmth. Also, baggage holding areas are usually unheated, and when temperatures get below freezing, an animal raised in a temperate climate may suffer when exposed to these conditions.

Before leaving for the airport, call the airline's 800 number or check their website for information about the weather both at the animal's destination and on route. Thunderstorms predicted along the way, especially around a major airline hub, can mean long delays and even canceled flights. When your animal is baggage, it is held up just like you would be, and sometimes even longer. If forecasted conditions seem likely to cause delay, consider postponing shipment. Nothing is more frustrating for both the shipper and person awaiting shipment than having the animal delayed or shipped by a different route. You can head off most of these problems by making alternate plans and discussing these with the person you're shipping to. Most people are more concerned about their animal arriving in good shape than about its arriving on a particular day and time. Be flexible, and make sure the weather is favorable before delivering your animal to the airline.

When trailering animals long distances, you need to consider the weather constantly. Most of us make long trips with animals during the hottest time of the year. Before committing yourself to a route, consider alternatives. Is one route likely to provide cooler temperatures than another? That route might be a bit longer, but if animals are comfortable, they can tolerate more time on the road. If the weather has been especially wet and flooding is a possibility, consider routes that don't cross major waterways or those less likely to flood.

Know ahead of time where there are facilities that you can use for housing your animals if weather or other conditions force you to stop. On a trip from California to Illinois with forty goats, our trip was interrupted by a tornado and a train wreck. If we hadn't mapped out the various fairgrounds and friends' houses along the route, we wouldn't have known where we could easily get our animals off the road, comfortable and safe. Since the train wreck closed east-west travel for two days, we had to have a place that could accommodate our needs. Having this information beforehand cut down on stress for all involved.

As you travel, listen to local radio stations for current and forecasted weather, and listen especially closely for conditions in areas you will be coming to in the next few days. If conditions seem problematic, make contingency plans. If may mean an alternate route or a delay, but these usually are better than being stuck with a trailer-load of screaming goats in the midst of bad, even dangerous, weather.

Without trusty equipment, you'll never make it.

Well before you leave on even a short trip with animals, check your car or truck and trailer for obvious and potential problems. Leave plenty of time to get any defect fixed. In the heat of the day, a flat tire or a broken belt can mean discomfort or even danger. A wheel that comes off, for example, can cause a crash and danger not just to you and your animals but to others on the road as well.

For short trips, do a walk-around before taking off. Make sure that tires are properly inflated, that all electrical and safety equipment (wipers, lights, hazard lights, horn, etc.) work, that there are no signs of oil or other fluid leakage. Fix any problem or postpone your trip until the equipment is shipshape.

For longer trips, have both the vehicle and trailer serviced well before you leave. Make sure that you are alerted to any potential problems, and fix them before you leave. Do the same preventive walk-around before you load up and take a final look before you pull out. The weight of twenty goats in the trailer may be just enough to cause a problem not evident before loading. Remember: problems rarely heal themselves, and breakdowns almost always occur on roads with no shoulders, no emergency call boxes, and no service stations.

For all trips, make sure you have the tools and parts you'll need (even if you don't know how to use them yourself) to fix common problems. Have a jack appropriate for your vehicle and loaded trailer. Also have blocking material, fluids that may need replacing, spare belts, filled spare tires for vehicle and trailer, and tools needed to make common repairs. If you're a repair novice, talk with your mechanic about what to bring, and if you travel a lot by yourself, consider taking an auto mechanics course offered at a local college.

And finally, a few tricks:

Having hauled large show strings across the country for a number of years, we developed a system that worked for our goats and us. Our goats arrived healthy and rested, ready to milk and show their best.

First, we followed the advice given above. We planned our route carefully, considering weather, road conditions, services along the way, etc. We made sure that our equipment was in the best shape possible.

We made our goats as comfortable as possible. We double-decked our trailer so that kids could be segregated from larger animals. They arrived far less stressed than if everyone traveled together. We bedded the trailer (top and bottom) heavily. The does rode on rubber mats plus bedding to lessen the road shock.

Goats had feed in front of them the whole trip. We attached flat hay feeders to the inside of the trailer (top and bottom) so that the animals who wanted feed could get it at any time. We replaced soiled feed and filled empty feeders as needed.

Whenever we stopped for fuel, we milked the does and watered everyone. On a long trip, this could mean milking five or six times a day. Sounds like a lot of work but once you get into it, it's easy. We'd string up a long tie chain along the trailer and hook up ten goats at intervals along the chain. One of us would strip each goat while the other watered. In the interest of time and space, some does got milked in the trailer. Twice a day we offered grain as we milked. Keeping goats milked out while they travel helps prevent udders from being banged in the trailer and helps maintain milk production under strenuous conditions. Offering water at every stop, keeps dehydration to a minimum and milk levels to a maximum.

For us, driving straight through, no matter how long the drive, worked best. Spreading a trip over a number of ten-hour days just seemed to prolong the agony of traveling, and the goats never rested well in foreign surroundings. So we would calculate how long the drive would take, how many drivers we would need, how much "down time" was needed. Our goal was to arrive as the gates opened so that the goats and we had the longest time to settle in, rest, and recuperate.

Having noticed that goats put in a new environment spend much of their first hours fighting each other and generally adding to the stress level of both goats and owners, we put a little something in their water to allow them to relax. Upon entering their pens, our goats drank some water, settled down for a short snooze and awakened refreshed and relaxed in their new surroundings.

For the first day, we tried to mess with them as little as possible, allowing them to recover from the stressful trip. This little bit of time was usually sufficient for them to gain back any production they had lost, and they were ready for the rigors of show life.

While many breeders have traveled more goats farther distances, few have been as successful as we have been in terms of minimizing equipment failure and maximizing the comfort and ease of travel for our animals (and us). Both of these translate into real benefits: not only are you and your goats rested and in excellent shape, but you may just have that little edge you need to win in strong competition.

Working Goats

We use the term "working goats" here to identify goats trained for packing and pulling. We acknowledge that milking goats are just as hard workers as pack and pulling goats, but this phrase seems an easy way to designate those trained for these special purposes, and those people who have pack and harness goats use it themselves.

In the past twenty years, we have seen a growing number of goats being raised for packing and pulling. The pack goat community owes a debt to early goat packers, including John Mionczynski, who, with his almost evangelical zeal, brought pack goats to prominence among the dairy goat community. Today, goat packing is popular, not just among those living close to wilderness and mountain areas, but a number of backpackers are finding their hikes much more enjoyable (and less strenuous) when accompanied by a pack goat.

As long as there have been domestic goats in the United States, there have probably been pulling or harness goats. If you look at old photographs of holiday parades, you are likely to see pictures of goat carts in line right behind the more familiar floats. So harness goats are not new, just our interest in them. Today, people are training goats to pull for fun and for work, and you can see goat carts today in holiday parades and on display at fairs.

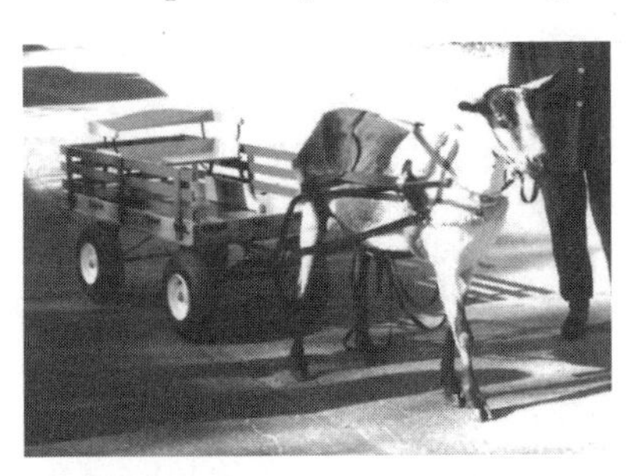

The nice thing about pack and pulling goats is that you can use any breed, as long as the animals are sturdy and well-built. There is at least one breeder of Cashmere goats who also uses her goats to pull. Pygmy goats trained to pull are a real attraction, and the only rule is not to load the cart with more weight than the goat can handle.

Those who are interested in learning more about pack goats should refer to breeder listings in national journals, and the American Harness Goat Association can give you information about training goats to pull.

Conditioning Your Pack Goat

The following article appeared in the Summer, 1997 issue of Goat Tracks: Journal of the Working Goat. *We wish to thank Ellen Herman, publisher, and Mary Wolf, author, for allowing us to reprint it here.*

Conditioning Your Pack Goat

by Mary Wolf

It's something we already know. Humans get out of shape. Not surprisingly, so do goats. Strapping on a heavy pack and hiking up a mountain when you've been barn or couch potatoes all winter makes for unhappy campers. Exercising with your goat regularly keeps both man and beast from sliding too far down the fitness ladder. It also provides quality time together. I believe goats enjoy our company just as much as we goat packers enjoy our caprine friends.

Because my goats are confined to a paddock area, I make the effort to take them on regular hikes. Not only do the goats stay fairly fit, they get to fine tune their manners, trail skills, and packing techniques.

Throughout the winter I take the goats for a hike once or twice a week. Because I live at the edge of a large metropolitan area, I must trailer them to nearby trailheads. Most of these are heavily traveled. I always saddle the goats. Not only does this make it easier to control the group, it keeps my saddling habits fine-tuned. We leave the trail head in a pack string and stay that way until well away from the "madding crowd." Then I turn them loose and we hike about five miles. There are a number of open space areas in the foothills plus a state park nearby that offer a variety of trails, water obstacles, and terrain. I alternate our trips to each area so we (the goats and me) don't become bored.

When spring arrives, I step up our outings to 2-4 times a week. The added trips are hill climbs to work legs and lungs. The panniers come out of storage. I pack them lightly at first, gradually increasing the weight by several pounds a week until we reach the weight each goat will carry on pack trips. I use local newspapers, bound and put in a plastic bag, to weight the panniers. It is stable and easy to add weight as needed. I also do a few "overnighters" close to home to resolve any problems with gear. Gear gremlins always manage to wreak havoc on my camping equipment during the winter. Easing into the pack season this way keeps stressful moments at a minimum. My goats, my gear, and I are well prepared and ready to enjoy our summer outings.

The best way to ensure a safe and successful packing trip is to get your goats out for a walk on a regular basis. Remember, if you miss a week here and there, don't go on a guilt trip. Just do the best you can and enjoy the time you do spend with your caprine pals. Goats, thank goodness, have flexible hours and can work around your schedule. They will be in shape and well trained. And so will you!

Happy trails....

Basic Training for Pack Goats

The following article appeared in the Summer, 1997 issue of Goat Tracks: Journal of the Working Goat. *We wish to thank Ellen Herman, publisher, and Rex and Terri Summerfield, authors, for allowing us to reprint it here.*

Basic Training for Pack Goats

by Rex and Terri Summerfield

The goats were large, each over 200 pounds. I could feel every ounce of it as I tried to heave them up into the back of a 4 x 4 pickup. I finally succeeded after much muttering about goats that wouldn't load by themselves. I might add here, that they were borrowed goats; my own were already in the truck. They had jumped in unaided and watched as I wrestled with the borrowed goats.

That incident happened several years ago and brought home the need to teach pack goats to lead and "load up." It's always a source of secret pride when my goats load in the truck with a quiet word and a pat on the tailgate, as onlookers watch, usually waiting for an excuse to laugh at you for using goats in the first place. Goats that handle easily impress people. It's that simple. It's also that aspect that brings new people to goat packing.

I feel that its every goat owner's responsibility to train their goats, not only in the basics (leading, loading, and packing) but also what I call camp manners. John Mionczynski's famous "goat management tool," the squirt gun, is almost indispensable for teaching camp manners. I usually start with "barn manners." No goats are allowed to stand on the fences, gates, etc. I'm sure transgressions take place in my absence but none are allowed when any of us are in the barn. It has been my experience that behaviors consistently discouraged when we are around eventually fade from the animal's list of habits.

Every goat is an individual, however, and some are very stubborn. A verbal command, usually "No," is spoken before the correction, and it doesn't take long before you can look at a goat and tell it "No," and it will stop what it is doing. Other goats in the area may also look nervously around to see who is going to get it, but at least you know they are paying attention. Just as you discourage bad habits, you should praise a goat for doing the things you like. This helps reinforce good habits.

One of the most commonly asked questions on goat training is when to start teaching a goat to lead. I like to start when they are only a few weeks old. The older they get, the harder they are to teach, not because older goats are any dumber, but simply because it's harder to pull them if they decide they want to stop. Some goats, especially younger ones, have short attention spans. It is better to do several short sessions than one long one. Remember to try to keep it fun for the goat. Be sure to include lots of praise. Some people coax the goat with treats, but I would rather have the goat think that the rope can't be defeated instead of learning that he has some control over being led. A 200+ pound wether that doesn't want to go is impossible to tug along behind you. If you already have a full size goat or acquire one that doesn't lead, then you may have to employ treats and extra coaxing to get the goat going.

One method I have used on older animals that refused to lead, or the extra stubborn ones, involves the use of an electronic dog training collar. I set the electrical stimulation down to low to start. First I pull the lead rope tight and if the goat doesn't give, I press the button and hold it for about one to two seconds. (Remember, it is on low setting.) I then let slack in the rope and let off on the button. Two or three times of that and the goat learns that a tight rope means a shock. It is important not to shock the goat if he gives any slack at all. In only a few short minutes the goat will immediately give slack when you pull the rope. After the goat has the system figured out, he may test it a time or two. All of the goats I have used this system on learned to lead perfectly in less than five minutes. Several lessons are required before the goat leads reliably without the collar. I might also mention that after the goat has the system figured out, I give a warning prior to the correction, just like the squirt gun. I use a light bump on the lead and a "heel " command prior to the shock. Even the most stubborn goats usually won't balk past the warning. It may sound a little harsh, but a goat that won't lead in an area requiring animals to be led is worthless. It also doesn't do much for goat packing in general, for onlookers to see a tug of war between you and your full grown goat.

Another skill a goat needs to learn is how to load into a pickup truck. I usually don't teach the goats to load until they are large enough to make the jump into the pickup unaided. Usually at a year old or so. I start by backing up to a hump or dirt bank, so there is only a six to twelve inch step into the truck. I pat the tailgate and tell the goat to "load up." I then lead the goat into the truck and praise it and give it a handful of grain. A few times in and out and the goat learns that if it goes in, it gets some grain. The trick after that is to keep the goat out of the truck until you give the command to load up.

After the goat is familiar with the routine, pull the truck ahead, increasing the distance from the ground to the tailgate. Continue this routine, until the goat is jumping in from normal ground level.

Some goats can be taught to load in a single lesson, others take a few lessons. If the goat doesn't get loaded regularly, you may have to review the lesson a couple of times. Since we give our goats some grain after each packing trip, it seems only natural that they get it after they load up.

As the TV commercial says, "A mind is a terrible thing to waste." Your goat is an intelligent animal, capable of learning lots of skills. If it is ignorant . . . well, it's not the goat's fault.

See ya in the brush!

An Introduction to Training Goats to Pull

This article is the introduction to the booklet, Training Goats to Pull a Cart, *published by and available from Caprine Supply.*

Most goats have sufficient strength and ease of temperament to make them excellent candidates for training to pull carts. Having harness goats can make work around your farm or homestead a little easier, it can help children develop excellent skills, and it can even keep the goats out of trouble (a goat occupied pulling can't be climbing through fences or getting into other mischief!).

Training a goat to pull can be an excellent 4-H project either for individuals or for a group of 4-Hers. Children learn and practice their general goat management skills including feeding, grooming, disbudding and castrating as they would with other projects.

However, training goats to pull requires tremendous patience: children learn to work step-by-step, working on small parts of the pulling process and settling for small victories. They learn that animals must completely master one step before moving on to the next step. They see that animals sometimes forget even the most obvious steps, making progress seem snail-like. Finally, they learn that their goal, a goat that can pull a cart, can be reached only after long, hard work on everyone's part.

Most of the skills they learn through a project such as this will form the basis for most of their future study and work.

An additional benefit of having pulling goats is one that comes to all goat owners, not just you. Goats have suffered from "bad PR." Everything we do that shows goats in a good light helps others see the worth of goats and goat products. However, usually our "good PR" efforts are limited to fairs and food shows. Having a goat trained to pull and an attractive cart or exhibit means you can show off goats at many community activities and can help the public see what fine animals goats really are.

This high quality wood and steel cart can carry 1,000 lbs., more than most goats can pull!

One last bit of advice: we talk a lot about selecting a goat to pull. I think it's equally important to make sure that you are someone who is comfortable with this kind of project. If you are someone who can tolerate what we call "delayed gratification," then this project is for you. Training a goat to pull is neither a quick nor easy process, and you may not see immediate results from your efforts. You must be happy with mastering small steps. However, in the end, the joy and fun of having your goat pull you or your children in a cart will be worth all the effort.

Goat Products: An Overview

When we first began raising goats in the early 1970's, there was not a large market for goat milk. In California and Arkansas, there were milk processing plants that produced powdered and evaporated milk. They bought grade B milk from local producers. In the Northern California area, Laurelwood Acres was producing fluid milk, and across the United States you'd find scattered Grade A dairies producing fluid goat milk for local markets. In the Midwest there was a cheese coop that produced various cheese products, and there was a group that marketed kids in the spring to the ethnic markets in the East.

The baby boomer generation with its tastes for goat cheese and health products, gourmet eateries, and ethnic food has encouraged an explosion in the development and production of goat-related food items.

If you visit almost any high-end restaurant in the United States, you'll have to go far to find a menu that **does not** feature goat cheese in its offerings, and today, many chain supermarkets in large urban areas feature goat milk and goat cheese alongside their regular cow milk products. And a few years ago, Cypress Grove's incredibly delicious Humboldt Fog cheese graced the cover of Dean and DeLuca's gourmet food catalog. Goat products have come a long way in thirty-five years.

Besides this explosion of interest in milk products, we are seeing an increase in demand for goat meat. This is being met by breeders of meat goats, who are working to produce a quality meat product for a market beyond the usual ethnic buyers. One slaughter house in Texas processes thousands of goat carcasses a week.

More and more goat owners are opening dairies, producing cheese, and raising meat goats. But it all starts in the home kitchen, and this section focuses on how to begin making and using goat products. If you are interested in developing a market for your goat products, you should first contact a producer of quality goat products in your area.

A simple cheese recipe

Slowly heat a gallon of milk to 185°F, using a stainless steel or enamelware pot. Do not use an aluminum kettle. Add 1/4 cup vinegar. Keep the temperature at 185°F, stirring the milk occasionally, for ten to fifteen minutes. At this point, a soft curd should form. Line a colander with cheesecloth. Pour the curd into the colander, and sprinkle the curd with salt. Tie the corners of the cloth together and hang it to drip for a few hours.

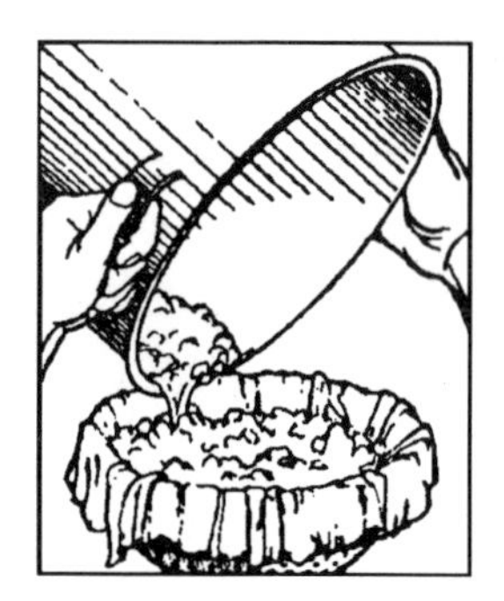

Add seasoning, such as dill, pepper or garlic, and refrigerate. Eat the cheese immediately, or keep it no more than a week in the refrigerator.

Rennet

Rennet is an enzyme that coagulates milk and causes cheese curds to form. For a soft cheese, you do not need rennet. You can use acid (lemon juice, vinegar, or lactic acid formed when the milk is inoculated with a special bacterial culture) to precipitate the curd. To form a stronger, firmer curd, you will need to use rennet.

Commercial rennet is an extract from the stomach lining of a milk-fed calf. Vegetable rennet, used to make vegetarian cheeses, is an extract from the mold Mucor Mieher. Rennet comes in liquid and tablet form. Liquid rennet is convenient and easy to measure accurately, but it must be refrigerated and kept away from prolonged exposure to light. Rennet tablets are easier to store, and they keep longer.

Starter Cultures

Packets of freeze-dried bacterial cultures provide the right "starter" for making cheese and yogurt. The bacteria in the starter turns lactose, which is milk sugar, into lactic acid, which helps form the curd in cheeses. The bacteria also contribute to the final flavor of the different cheeses.

Instructions for incubating the cultures are printed on each packet. If you work with your starter cultures carefully, you can keep cultures going practically forever, and you may not need to buy cultures very often. For more information about general cheesemaking, and starter cultures specifically, buy a cheesemaking book that features simple recipes and easy-to-follow directions.

Hint: Why Cheesemaking Fails

Most cheesemaking fails because of unclean equipment or contaminated milk. Here are a few hints to help make cheesemaking a success. Use only glass, stainless steel, or enamel-lined equipment. Also check that there are no chips or cracks on equipment surfaces. Go over your milking routine to ensure clean milk. Before you begin making cheese, check that all cheesemaking equipment is clean and sterilized. Before assuming that your culture is at fault, be sure to go over your milk handling and equipment cleaning routines.

Making Butter

You can make butter without a costly cream separator or mechanical churn. Here is how to do it:

Put fresh milk in wide-mouth quart jars in the refrigerator. The next day, skim off the little bit of cream that has risen to the top. Do this until you have enough cream to make your butter. You can freeze the cream to save it until you have enough. Obviously, by using a cream separator, you would get more cream faster.

When you have a pint of cream, put it in a quart jar and let it come to room temperature. Then shake the jar for about 15 minutes. When the butter forms, it gets to a "gloppy" stage and makes a thud instead of a sloshing sound. Next, pour the thin whey off. Rinse the butter with cool water two or three times, until the water runs off clean. In hot weather, refrigerate the butter before you rinse it.

Finally, put the butter in a bowl, salt to taste (if you want salted butter), and work it with a spoon to mix in the salt and get the rest of the whey out. Pour off any whey that has accumulated. The butter made from goat milk is white. If you want your butter to be yellow, add 10 to 12 drops of butter color to the pint of cream before you begin shaking it.

Making yogurt

Either buy a yogurt culture or use an unflavored yogurt with live culture (such as Dannon) as your starter.

Calves "Market" Milk

Sterilize all the equipment with boiling water just before you start.

Heat the milk to scalding and then cool it down to 110°F. Put the milk into small jars or cups. Add one tablespoon of yogurt to each pint.

Incubate the jars at 110°F for 3-6 hours or overnight. To do this, use a commercial yogurt maker or put your jars of yogurt on a heating pad (warm setting) and cover them with towels. You can also set the jars in a gas oven right above the pilot light.

Hints: Calves "Market" Milk

"I raised four Holstein calves to six months of age and sold them as baby beef. They started at a gallon of goat milk each per day, and they were up to 1 1/2 gallons each per day by five months. They also got free-choice hay (I fed them hay recycled from the goat pen. Remember that goats are much more finicky than calves about their hay.) They also got grain, one pound a day each to begin. At six months, they were getting four pounds each per day.

"I sold them direct to customers. I had them butchered at a local locker and charged $1.50 per pound for the carcass (this was in the early 1980's). The beef was extremely tender and tasty, and I already have orders for more next year.

"If I had bought grain in bulk, my expenses would have been less, but I paid all my expenses and made a tidy profit—without the trouble of trying to run a dairy.

"A word of warning: you must like calves to do this. And I was lucky. My calves were healthy, so I didn't lose any (a good thing since they were expensive to buy). If I raise 20 calves next year, I'll have an excellent "market" for my goat milk."

Goat on the Grill

When we were traveling in Mexico City years ago, we went to a restaurant that served only goat. In fact, the neon sign in the window showed a goat, and we could choose from delicacies including head, hind or front quarters. Needless to say, we opted for legs, not head, and the meal was both memorable and delicious.

At our house, our favorite way to fix goat is on the grill. We use any size goat, but the younger the goat, the tenderer it will be, and it takes less time is need to cook it. With young goats (under four months), we split or quarter them. For larger animals, we use a leg roast.

Our key to grilling goat is the marinade. We have tried fancy ones but keep coming back to plain ol' Italian dressing to which we add garlic and herbs (bay leaf and others of our choice). We put the meat, marinade and seasonings in a plastic bag, coating the meat completely, and let it to sit at least overnight in the refrigerator.

We start our fire (or gas grill) and allow the coals to burn down. When the coals are just right, we place the meat on the grill and allow it to cook, with the cover down, until the meat pulls away from the bone. Timing will depend on the age of the animal and the success of the marinade.

When the meat is tender (it should fall off the bones), we pull it from the bones, and serve it with our favorite BBQ sauce (we're hooked on a local Kansas City product called Boardroom at the moment!) on the side. To reheat, we place the meat in a ovenproof dish with a little broth, cover, and set it in a 350°F oven until it is heated through. We try not to let the meat dry out.

Cooking—The Cowboy Way

This article was originally published in the September, 1997, issue of the Goat Rancher *magazine. We wish to thank Terry Hankins, editor and publisher, for allowing us to reprint an edited version of it here.*

Cooking—The Cowboy Way

by Rusty Fleming

The cooking's done, the results are in, the champion crowned. Jay Butler of Snyder, Texas, is the 1997 World Champion Boer Goat Cook at the competition held in Loraine, Texas.

If you have ever wondered how many ways there are to cook cabrito (Spanish for little goat), all you have to do is attend a function like this one. Jay won't share all his secrets with us (his recipes were worth $1,000 and three trophy belt buckles), but cooking good cabrito doesn't have to be a deep, dark secret. Like many other things in life, it's best to keep it—the cooking—simple.

The first thing you need is, obviously, cabrito or chevon or goat meat. If you purchase the meat from a meat market or a packing plant, much of the work will have been done for you. If you are on your own, dress the goat like a deer. Truss it up by the hind feet, get the hide off, and then finish dressing the carcass. If you let any hair touch the meat, take a rag soaked in vinegar and wipe the meat down, letting the vinegar dry as the meat cools. This will help prevent tainted meat or off-flavor caused by the hair. It works well when butchering Angora goats where it's difficult not to contaminate the meat with their hairy coat.

After the meat has cooled, cut up the carcass. If the goat weighs 20 pounds or less, you can split the carcass down the middle, or do as the Mexicans do, cutting through the backbone where the first rib meets the backbone ahead of the hind saddle (loin), yielding a front half and a back half.

With goats larger than 20 pounds, first remove the shoulders, then the ribs. The ribs are removed most easily by using a mitre or dehorning saw, cutting the rib bones about one inch from the outside edge of the loin. Then cut the neck from the backbone and the backbone from the front of the hind saddle. Finish up by splitting the pelvis, using the saw, yielding the two hams or hind legs. This will give you nine pieces of cabrito which you can cook all at once or freeze some for later use.

Our preferred method for cooking goat is to barbecue it, or more accurately to smoke it. We use a firebox type pit which has the fire on one end of the pit, allowing the heat and smoke to pass around and through the meat.

We also cook exclusively with mesquite wood. Lots of people complain that they don't like mesquite since they say it leaves a rank, oily taste to the meat. And they're right. That will happen every time you use green wood. It will also happen if you use green hickory, oak, apple, pecan, peach, or any other kind of hardwood. So, for your perfect barbecue, be sure to use only dry, seasoned wood or charcoal.

If for any reason you feel the meat will not be tender, then take the time to parboil it. Take a large pot of water, bring it to a boil and put the meat in. Allow the water to come back to the boil and remove the meat. Parboil the meat for no more than five minutes.

Next, I rub the meat with a mixture of equal parts salt, black pepper, garlic powder, and a little paprika (for color). I place the meat on the fire for two to three hours until the meat sets. Setting is when the meat firms up and has a beautiful red color from the smoke and seasonings.

At this point I remove the meat from the fire and wrap it in heavy aluminum foil. I then continue cooking it for another couple of hours. The time will vary from goat to goat, so remember that your objective is to cook it until the meat is "fork" tender. When it's done, take the meat off the fire. You can now either strip the meat from the bone or leave it alone.

You can serve the cabrito with your favorite barbecue sauce, but let the sauce complement the meat, not smother it. You can also serve goat with green chili sauce. To make this, combine chopped green chilis, chicken broth, salt, garlic, and a touch of cumin, and cook the mixture until it begins breaking down and looks like lumpy gravy. For a milder green chili sauce, add two tablespoons of olive oil or margarine to the sauce as it cooks. To thicken the sauce, add a little cornstarch.

If you don't have a fancy smoker or barbecue pit, don't despair. You can use any kind of backyard cooker, or even your kitchen oven and still have super cabrito.

A word of warning: Every time you have the uncontrollable urge to take a peek at your great barbecue while it's cooking, you'll have lost fifteen minutes of heat and humidity, which will mean fifteen more minutes of cooking time.

Cooking—The Cowboy Way

Cooking goat isn't hard. With a little practice you can become a superb goat cook. And if you have any questions or run into a problem cooking goat, feel free to call me.

Equipment Checklist for Novice Goat Owners

Getting started with goats requires more than just bringing them home and putting them out to pasture. To keep animals healthy and well-fed, you will need some basic equipment. Besides the equipment listed below, be sure you have a good goat management book so you can look up answers to questions you will have. Finally, keep two important numbers by the phone: your veterinarian's and your local "goat expert's."

Feeding (kids):

- Nipples (pop bottle, Pritchard, Caprine nipples) ☐
- Caprine feeding outfit ☐
- Milk replacer ☐

Feeding (does, bucks, growing kids):

- Hay ☐
- Grain ☐
- Minerals ☐
- Feed pans ☐
- Hay feeders ☐
- Mineral feeders ☐
- Grain scoop ☐
- Water heaters (if needed) ☐

Animal Identification:

- Tattoo outfit with digits ☐
- Tattoo ink ☐
- Ear tags (Boer/hair goats) ☐
- Number tags and chain ☐
- Leg bands ☐

Kidding:

- Iodine ☐
- Towels ☐
- OB leg snare/kid puller ☐
- Navel clips ☐
- Surgical gloves ☐
- ID neck bands ☐
- Colostrum replacer ☐
- Goat coats (kid size) ☐

General Equipment:

- Disbudding iron ☐
- Kid holding box ☐
- Hoof trimmer ☐
- Collars (plastic/nylon) ☐
- Leads ☐
- Weigh tape ☐
- Goat coats (adult size) ☐
- Fencing equipment ☐
- Snaps, swivels, clamps, hooks ☐

Milking Equipment:

- Milk stand, restraining device ☐
- Bucket ☐
- Strip cup ☐
- Udder wash ☐
- Teat dip ☐
- Teat dipper/sprayer ☐
- Udder ointment ☐
- CMT test kit ☐
- Dairy towels ☐

Milk Handling Equipment:

- Milk tote ☐
- Milk strainer ☐
- Milk filters ☐
- Brushes ☐
- Detergent ☐
- Scour pads ☐
- Home pasteurizer ☐
- Milk thermometer ☐

Checklist for the Goat Medicine Chest

In our barn we have a kitchen cabinet that we use for storing medicines and equipment we need for treating routine and unexpected illnesses and emergencies. Below is the list of items we keep stocked there. You might want to copy it and tack it up on the door of your cabinet. Then, check your list periodically to see what items you need to replace.

- A good goat health book ☐
- Veterinary thermometer ☐
- 3 cc plastic syringes ☐
- 6 cc plastic syringes ☐
- 12 cc plastic syringes ☐
- 35 cc plastic syringes ☐
- Weak kid syringe (for tube feeding) ☐
- Needles (1" and 1/2") ☐
- 10, 30, 50 cc drenching syringes ☐
- Balling guns (small, large for pills) ☐
- Surgical scissors ☐
- Scalpel, disposable blades ☐
- Surgical gloves ☐
- Disposable scalpels ☐
- Glentle iodine spray (for wounds) ☐
- Tincture of iodine (dipping navels) ☐
- Blu-Kote (for wounds) ☐
- Blood stop powder ☐
- Propylene glycol (for ketosis) ☐
- Nutri-Drench/Power Punch (for stress) ☐
- Probiotic powder or paste ☐
- Vet Rx (for respiratory ailments) ☐
- Ketocheck (to diagnose ketosis) ☐
- Electrolyte powder (for stress) ☐
- Revitilyte Plus supplement ☐
- Amprolium (for coccidia) ☐
- Neomycin oral treatment (for scours) ☐
- Multi+Care Ointment ☐
- Vetwrap ☐

In the refrigerator:

- Procaine Penicillin G ☐
- Enterotoxemia vaccine ☐
- Tetanus toxoid ☐
- Tetanus antitoxin ☐

Checklist for Showing

Does this sound familiar? You have driven 400 miles to a show but upon your arrival you discover that you left the health papers on the kitchen table or the grain back in the barn. Either of these can mean a wasted trip or unnecessary expenses or both. To help avoid these panicked moments, use the following checklist to help you pack your tack box.

Feeding Equipment (kids, adults):
- Hay feeders ☐
- Grain pans or feeders ☐
- Water buckets ☐
- Caprine feeder ☐
- Nipples and bottles ☐
- Snaps or spring hooks ☐
- Grain scoop ☐
- Hay hook ☐
- Salt, salt holder ☐
- Water additives (vinegar, etc.) ☐

Milking Equipment:
- Milk stand ☐
- Milk bucket(s) ☐
- Milking machine ☐
- Teat dipper ☐
- Teat dip ☐
- Double end leads ☐

Grooming Equipment:
- Electric clippers ☐
- Cordless clippers ☐
- Blades for clippers ☐
- Clipper spray ☐
- Clipper grease ☐
- Grooming brush ☐
- Hoof trimmers ☐
- Hoof plane ☐
- Animal shampoo ☐
- Showsheen ☐
- Udder cream ☐
- Towels ☐

Miscellaneous:
- Registration papers ☐
- Health papers ☐
- Electrical extension cords ☐
- Goat coats, t-shirts ☐
- Ringside tie chains ☐
- Tattoo outfit with digits ☐
- Rapid water heater ☐
- Fence tools, pliers, hammer ☐
- Tarpaulin (windbreak or sunshade) ☐

Health Items:
- Thermometer ☐
- Kaopectate or Pepto-Bismol ☐
- Antibiotics, vitamins ☐
- Syringes, needles ☐
- Vetwrap ☐
- Power Punch, Nutri-Drench ☐
- Probiotic powder or paste ☐
- Electrolyte Powder ☐

Clean-Up Equipment:
- Broom, rake, pitchfork ☐

Feed (often available at show)
- Hay ☐
- Straw, other bedding ☐
- Grain ☐
- Minerals, supplements ☐

Exhibitor's Needs:
- Sleeping bag, pillow, cot ☐
- Food, cooler, thermos ☐
- Clothes (including show) ☐
- Towels ☐

Associations, Registries, and Goat Clubs

Alpines International
8194 Golf Link Road
Hilmar, CA 95324
www.alpinesinternationalclub.com

American Boer Goat Assn.
1207 S. Bryant Blvd., Suite C
San Angelo, TX 76903
325-486-2242 www.abga.org

American Dairy Goat Assn.
P. O. Box 865, Spindale, NC 28160
828-286-3801 www.adga.org

American Goat Society
735 Oakridge Lane
Pipe Creek, TX 78063
830-535-4247
www.americangoatsociety.com

American Kiko Goat Assn.
Lez'le Denson, Secretary
311 Apache Springs Road
Briggs, TX 78608
www.kikogoats.com

American LaMancha Club
11075 Old Oregon Trail
Redding, CA 96003
530-241-4214
www.lamanchas.com

American Meat Goat Assn.
P. O. Box 676, Sonora, TX 76950
325-387-6100
www.meatgoats.com

American Nigerian Dwarf Dairy Assn.
Memberships to: Barbara Brown
1510 Bird Road
Independence, KY 41051
www.andda.org

Canadian Meat Goat Assn.
Box 61
Annaheim, Saskatchewan
S0K 0G0
306-598-4322
www.canadianmeatgoat.com

Canadian Goat Society
2417 Holly Lane
Ottawa, Ontario K1V 0M7
613-731-9894
www.goats.ca

Golden Guernsey Breeders of America
5180 San Felipe Road
Hollister, CA 95023
408-396-6439
www.guerseygoats.com

International Boer Goat Assn.
P. O. Box 1045
Whitewright, TX 75491
877-402-4242
www.intlboergoat.org

International Fainting Goat Assn.
2455 Deanburg Road
Pinson, TN 38366
www.faintinggoat.com

International Sable Breeders's Assn.
http://sabledairygoats.com/

International Nubian Breeders Assn.
Caroline Lawson, Secretary/Treasurer
5124 Fm 1940
Franklin, TX 77856
979-828-4158
www. i-n-b-a.org

Kinder Goat Breeders Assn.
P. O. Box 1575. Snohomish, WA 98291
www.kindergoats.com

Miniature Dairy Goat Assn.
P. O. Box 7244
Kennewick, WA 99336
509-591-4256
www.miniaturedairygoats.com

National Pygmy Goat Assn.
1932 149th Ave. SE
Snohomish, WA 98290
425-334-6506
www.npga-pygmy.com

National Saanen Breeders Assn.
PO Box 916
Santa Cruz, NM 87567
http://nationalsaanenbreeders.com

National Toggenburg Club
Memberships to:
1156 East 4100 North
Buhl, ID 83316

www.nationaltoggclub.org
Nigerian Dwarf Goat Assn.
8370 W. Abbey Ln.
Wilhoit, AZ 86332
928-445-3423
www.ndga.org

Oberhasli Breeders of America
Elise Anderston, Secretary/Treasurer
1035 Bardin Road
Palatka, FL 32177
http://oberhasli.net

Pygora Breeders Association
538 Lamson Rd., Lysander, NY 13027
315-678-2812
www.pygoragoats.org/

United States Boer Goat Assn.
P. O. Box 663, Spicewood, TX 78669
1-866-66U-SBGA (1-886-668-7242)
www.usbga.org

Magazines

Countryside & Small Stock Journal
145 Industrial Drive
Medford, WI 5445
800-551-5691
http://www.countrysidemag.com

Dairy Goat Journal
145 Industrial Ave.
Medford, WI 54451
800-551-5691
www.dairygoatjournal.com

United Caprine News
P. O. Box 328
Crowley, TX 76036
817-297-3411
www.unitedcaprinenews.com

Goat Rancher
225 Hankins Rd.
Sarah, MS 38665
888-562-9529
www.goatrancher.com

The Goat Magazine
P. O. Box 2694
San Angelo, TX 76902
325-653-5438
http://goatmagazine.info

Goatworld.com (online magazine)
719-267-4279
509-448-8758

Goat Tracks
558 Park Ave.
Logan, UT 84321
www.goattracksmagazine.com

Mother Earth News
1503 SW 42nd St.
Topeka, Kansas 66609-1265
http://www.motherearthnews.com

Ruminations
P. O. Box 859
Ashburnham, MA 01430
www.smallfarmgoat.com

Laboratories that Provide Testing for CAE Virus

California Animal Health & Food Safety Lab System
West Health Sciences Drive
Davis, Calif. 95617-1770
530-752-8700
cahfs.ucdavis.edu

Animal Health Diagnostic Center
Cornell University
College of Veterinary Medicine
P. O. Box 5786
Ithaca, New York 14853-5786
diaglab.vet.cornell.edu

Pan American Veterinary Labs
166 Brushy Creek Tr.
Hutto, TX 78634
800-856-9655
www.pavlab.com

Veterinary Diagnostic Laboratories
University of Minnesota
College of Veterinary Medicine
1333 Gortner Ave.
St Paul, Minnesota 55108-1093
800-605-8787
www.vdl.umn.edu

Washington Animal Disease Diagnostic Laboratory
College of Veterinary Medicine
Washington State University
PO Box 647034
Pullman, Washington 99164-7010
509-335-9696
www.vetmed.wsu.edu

We have tried to provide accurate information about phone numbers, addresses, and websites. Unfortunately, numbers and contact people for various clubs and associations change. We suggest that you check the Internet or directory assistance to check this information as time goes on.

Resources: Web Sites of Interest

Alpines International Club
http://www.alpinesinternationalclub.com

American Boer Goat Association
http://www.abga.org

American Dairy Goat Association
http://adga.org

American Goat Society
http://www.americangoatsociety.com

American Kiko Goat Association
http://www.kikogoats.com

American LaMancha Club
http://www.lamanchas.com

American Minor Breed Conservancy
http://www.albc-usa.org

American Nigerian Dwarf Dairy Association
http://www.andda.org

Canadian Goat Society
http://www.goats.ca

Cashmere Goat Information
http://www.capcas.com

DNA Testing for Goats
http://www.gtg.com.au/AnimalDNATesting/
index.asp?menuid=080.200
http://www.biogeneticservices.com/animalgenotyping.htm
http://www.vgl.ucdavis.edu/forms/Goat_ID_Form.pdf

Embryo Transfers
http://www.creeksideanimalclinic.com

International Boer Goat Association, Inc.
http://www.intlboergoat.org

International Fainting Goat Association
http://www.faintinggoat.com

International Kiko Goat Association, Inc.
http://www.theikga.org

International Nubian Breeders Association
http://www.i-n-b-a.org

International Sable Breeders Association
http://sabledairygoats.com

General Goat Information Websites
http://www.greenspun.com/bboard/q-and-a.tcl?topic=Dairygoat
http://www.aces.edu/pubs/docs/indexes/unpas.php#small
http://www.vetmed.ucdavis.edu/vetext/INF-GO_CarePrax2000.pdf
http://animalscience.tamu.edu/academics/sheep-goats/index.htm
http://www.goatworld.com
http://www.goatworld.com/911
http://www.goatbiology.com/parasites.html
http://www.farminfo.org/livestock/goats.htm
http://www.sheepandgoat.com
http://www.cals.ncsu.edu/an_sci/extension/animal/meatgoat/ahgoats_index.html
http://www.boergoats.com/clean/coverpage.php
http://sfc.ucdavis.edu/goatmeatpub.pdf
http://sheepgoatmarketing.info
http://www.goattalk.com/forum
http://www.hobbyfarms.com
http://www2.luresext.edu/goats/index.htm
http://aipl.arsusda.gov/eval.htm

Kinder Goat Breeders Association
http://kindergoats.com

Meat Goat Production article
http://www.aces.edu/pubs/docs/U/UNP-0104

National Pygmy Goat Association
http://www.npga-pygmy.com

National Saanen Breeders Association
http://nationalsaanenbreeders.com/

National Toggenburg Club
http://nationaltoggclub.org

Nigerian Dwarf Goat Association
http://www.ndga.org

Resources: Web Sites of Interest

Oberhasli Breeders of America
http://oberhasli.net

Pedigree Internationl LC (registers Genemasters)
http://www.pedigreeinternational.com

Pygora Breeders Association
http://www.pygoragoats.org

United Caprine News
http://www.unitedcaprinenews.com

United States Boer Goat Association
http://usbga.org

USDA Scrapie Information
http://www.aphis.usda.gov/animal_health/animal_diseases/scrapie

Index

Index

Index

Index